OPTIMAL STRATEGIES IN SPORTS

Studies in Management Science and Systems

Editor

Burton V. Dean

Department of Operations Research
Case Western Reserve University
Cleveland, Ohio

VOLUME 5

NORTH-HOLLAND PUBLISHING COMPANY
AMSTERDAM · NEW YORK · OXFORD

Optimal Strategies in Sports

Edited by

Shaul P. Ladany
Ben-Gurion University, Israel

and

Robert E. Machol
Northwestern University, U.S.A.

1977

NORTH-HOLLAND PUBLISHING COMPANY
AMSTERDAM • NEW YORK • OXFORD

North-Holland ISBN: 0 7204 0528 9

Published by:

North-Holland Publishing Company
Amsterdam • New York • Oxford

GV
706.8
.O 67

Sole distributors for the U.S.A. and Canada:

Elsevier North-Holland, Inc.
52 Vanderbilt Avenue, New York, N.Y. 10017

Library of Congress Cataloging in Publication Data
Main entry under title:

Optimal strategies in sports.

    (Studies in management science and systems ; v. 5)
    Bibliography: p.
    Includes index.
    1. Sports--Mathematical models--Addresses, essays, lectures. 2. System analysis--Addresses, essays, lectures. 3. Mathematical optimization--Addresses, essays, lectures. I. Ladany, Shaul P. II. Machol, Robert Engel, 1917-
GV706.8.O67   796'.01'84   77-471
ISBN 0-7204-0528-9

Printed in the Netherlands

TABLE OF CONTENTS

Table of Contents vii

Contributors

Richard Bellman, Departments of Electrical Engineering, Mathematics, and Medicine, University of Southern California, Los Angeles, CA

Bernard Benjamin, Professor of Actuarial Science, The City University, London, England

Paul D. Berger, School of Management, Boston University, Boston, MA

Paul Bratley, Département d'informatique, Université de Montréal, Montreal, Canada

M. N. Brearley, RAAF Academy, Victoria, Australia

William O. Cain, Jr., School of Business Administration, Emory University, Atlanta, GA

Virgil Carter, Chicago Bears Football Team, Chicago, IL

Earnshaw Cook, 3908 N. Charles St., Baltimore, MD 21218

D. A. D'Esopo, Stanford Research Institute, Menlo Park, CA

Arthur J. Francia, Accounting Department, University of Houston, Houston, TX

R. Allan Freeze, University of British Columbia, Vancouver, Canada

Shelby J. Haberman, Department of Statistics, University of Chicago, Chicago, IL

Dennis R. Heffley, Department of Economics, University of Connecticut, Storrs, CT

Ronald A. Howard, Department of Engineering-Economic Systems, Stanford University, Stanford, CA

John W. Humes, Baruch College, The City University of New York, New York, NY

Joseph B. Keller, Courant Institute of Mathematical Sciences, New York University, New York, NY

Shaul P. Ladany, Department of Industrial Engineering and Management Engineering, Ben Gurion University of the Negev, Israel

B. Lefkowitz, Allstate Research Center, Menlo Park, CA

G. R. Lindsey, Operational Research and Analysis Establishment, Department of National Defence, Ottawa, Canada

Robert E. Machol, Graduate School of Management, Northwestern University, Evanston, IL

Allen J. Michel, School of Management, Boston University, Boston, MA

Carl R. Mitchell, Scientific Sports Service, Boston, MA

James P. Monahan, School of Management, Boston University, Boston, MA

Carl Morris, The Rand Corporation, Santa Monica, CA

Frederick Mosteller, Department of Statistics, Harvard University, Cambridge, MA

Peter Palmer, System Development Corporation, Lexington, MA

Arthur V. Peterson, Jr., Department of Biostatistics, University of Washington, Seattle, WA

Richard Pollard, London School of Hygiene and Tropical Medicine, London, England

Stephen M. Pollock, Department of Industrial and Operations Engineering, University of Michigan, Ann Arbor, MI

Richard C. Porter, Department of Economics, The University of Michigan, Ann Arbor, MI

Bertram Price, Graduate School of Business Administration, New York University, New York, NY

J. Gerry Purdy, Samaritan Health Service, Phoenix, AZ

Ambar G. Rao, Graduate School of Business Administration, New York University, New York, NY

Charles Reep, Soccer Consultant, Torpoint, Cornwall, England

Frank Ryan, Committee on House Administration, U.S. House of Representatives, Washington, D.C.

Frank Scheid, Department of Mathematics, Boston University, Boston, MA

William Simon, School of Medicine and Dentistry, University of Rochester, Rochester, NY

John H. Smith, Professor Emeritus of Mathematics and Statistics, American University, Washington, D.C.

Georghios P. Sphicas, Baruch College, The City University of New York, New York, NY

Robert H. Strawser, Department of Accounting, Texas A&M University, College Station, TX

Richard E. Trueman, Professor of Management Science, California State University, Northridge, CA

INTRODUCTION

by

Shaul P. Ladany and Robert E. Machol

This is a book about the application of quantitative methods and systems analysis to sports events. The mathematics is mostly not very complicated, and the intelligent layman should be able to read most of this book
with profit and enjoyment. Nonetheless it is extremely incisive, and most
of the articles come up with specific recommendations for managers and
other decision makers in sports competition.

The question naturally arises, what in the world do quantitative techniques have to do with sports? We can imagine Connie Mack or Casey Stengel
turning over in their graves at the suggestion that some theoretician could
tell them how baseball ought to be played. And yet decisions which were
routinely made by Mack and Stengel, and which are being routinely made by
today's managers, are obviously wrong--obvious, that is, to any reader of
this book. Of course Mack and Stengel did a lot of things correctly--else
they would not have won so often; but they still employed the sacrifice
bunt and intentional walk too often, for example. This is proven in articles
in this book by Lindsey, who uses primarily empirical data backed by theory;
by Trueman, who uses primarily theoretical arguments backed by data; and by
Cook, who uses a computer to simulate thousands of baseball games. The intentional walk is particularly bad in the early innings, and even in late
innings it could only be useful if one were passing a strong batter to get
to a weak batter, and then only if one were sure that a pinch-hitter would
not be substituted for the weak one. The reasons for this conclusion can
be readily seen in the quantitative analyses employed by these authors;
the intentional walk puts an extra man on base who may score, and the increase in probability of a double play is so miniscule that it simply does
not compensate except in extraordinary circumstances. (Of course in the
bottom of the ninth with the score tied the intentional walk can do no
harm.)

Thus managers of sports teams should read this book, as should analysts
and serious students. But the primary intended audience is the intelligent
layman and sports fan. The latter will probably not be able to read all of
the text of all of the articles, although the following chapter, entitled
"A Word to the Nonmathematician" does attempt to explain a few of the most
common technical terms used in this book, such as "mean" and "variance".
In most of the articles there is no advanced mathematics; and in the few
exceptions, the layman can read around the math to get to the meat of the
article. For example, Brearley uses heavy mathematics (vector calculus),
which the layman can skip to get to his conclusions about whether Beamon's
miraculous jump at Mexico City was affected by the altitude. Similarly,
Keller uses advanced mathematics (calculus of variations) to derive the
optimal velocity with which a race should be run, but the layman can skip
this and still note the nature of Keller's conclusions and the remarkably
good fit of his theory to the actual data of recorded times at different
distances; and he can understand Keller's logic, contravening the conventional wisdom--which, for example, dictates that the last quarter of a mile
race should be run faster than the second and third quarters. These articles

are the exceptions; in most of the articles, the math is no more complicated
than elementary statistics.

The kinds of technical tools used can be visualized from the table
shown here (analysts would call such a table a "matrix"), where the author
(or first author in case of co-authored articles) is listed under a column
corresponding to the sport and a row corresponding to the tool. Almost
every article uses probability and statistics; only a very few use advanced
techniques.

	Baseball	Football	Track and Field	Miscellaneous
Probability and Statistics	Bellman Cook D'Esopo Freeze Howard Lindsey Mitchell Palmer Peterson Simon Smith Trueman	Carter Haberman Mosteller Pollard Porter	Ladany Pollard	Bratley Monahan Morris Pollard Pollock Price Scheid
Matrix Algebra	Bellman			
Calculus			Brearley Keller	Pollock
Optimization Theory	Bellman Howard Cain		Heffley Ladany	Machol
Computers	Cain Cook Freeze Howard Mitchell Peterson Purdy Trueman	Purdy Ryan	Purdy	Price Purdy

This book is part of a two-pronged effort. The other book, under the
same co-editorship, is entitled "Management Science in Sports" and was published a few months earlier by North-Holland as a special issue of the
scholarly journal Management Science. It consisted of 16 articles and a
short note, each reporting on original research not previously published,
and mostly written for the professional in operations research/management
science. Thus, that book (its table of contents is presented on page 207)
may be more difficult for the layman than this one.

We believe that managers are eventually going to read these articles and improve their strategies, but in the interim, the reader of this book will know how the game ought to be played and why. He will also understand better how a fan should react to the effects of chance on sports competition. Philip Morrison put this very well in reviewing the book by Cook and Garner (that book is summarized in Cook's article in this volume): "The most interesting feature of the book is an account of the interplay of chance and cause. The World Series are so short, and played between teams so nearly equal, that their results and even their duration are nearly pure chance. Batters obviously differ in skill, but most of their slumps and streaks cannot be distinguished from randomness. Mickey Mantle's good and bad years agree with chance. The greatest hitters were, Cook and Garner say, Ty Cobb, Ruth and Ted Williams; the correlation between scoring indexes and general informed judgment is excellent. Nonetheless, the contrast between the run of sheer probability and the overheated effort to understand causally the inner reason for each loss, each "streak," each day-by-day fluctuation is incontrovertible. Most real events of the game are simply too complex to interpret and ought in all reason to be handled with statistics, that is, patiently."*

Sometimes the strategies suggested in this book have not been adopted simply because nobody has thought of them before. Carter and Machol published their original article (it has been rewritten and improved here) in 1971, suggesting that time-outs be called later in the last two minutes, and that coffin-corner punts be attempted more often. Both of these strategies are now more frequent.

Sometimes strategies are not adopted out of sheer conservatism. The Australians introduced a new way of placing oars in an eight-oar shell in the early 1960's, as explained by Brearley, but in the 1976 Olympics nobody except the Australians had adopted this obvious improvement. Perhaps it will have to wait until the Australians win an international race, and then everyone else will copy them.

Sometimes strategies are not adapted for reasons unconnected with winning the competition. For example, Lindsey and others prove that if there are two outs and there is a man on third base, he should attempt to steal home, even if his probability of success is as small as 1/3. The reasons are obvious. The chances that the next at-bat will be an out are better than two out of three, and in this case the man will die on third. Yet frequently we have runners whose probability of stealing home safely is at least one third, and who are not sent home by their managers. The reasons are intriguing. If a manager sends three runners home from third, and if one of them scores and two of them are out, he may end up winning more ball games, but he will also end up looking more stupid, and managers who end up looking stupid are not likely to keep their jobs. Thus, in good systems analysis, as described in this book, it is essential that the objective function be kept clearly in mind.

A similar example comes from punting in football. A kicker who booms it into the end zone instead of putting it out of bounds on the 10, or who

---

sends long line drives instead of hanging it high, may win less ball games, but his average (yards per punt) is going to be bigger. If a manager uses this average to the exclusion of everything else in salary negotiations, he's going to encourage the wrong kind of behavior. Similar remarks apply to many other statistics used in evaluating players. The win-loss record of a pitcher is clearly affected by many other things than his own capabilities and performance, and while the earned-run average is better, it too is an incomplete description. If the batting average captures the capabilities of hitters, then it clearly captures just as well the capability of the pitcher in the batter-pitcher confrontation, but no one seems to have bothered computing the batting averages of batters facing a particular pitcher. Part of the reason is that the batting average itself is not a very good measure. Since an extra-base hit is clearly better than a single, the slugging average is better than the batting average (Lindsey shows that this can be improved by weighting doubles, triples and home runs at 1.8, 2.3, and 3.1, respectively, rather than 2, 3, and 4). Obviously walks are better than nothing (Lindsey says a walk is worth 7/10 as much as a single), and there ought to be some penalty for grounding into a double play--there may be nothing wrong with hitting a hard groundball, but the man who usually beats the relay is a better hitter than the man who usually doesn't. Thus, anybody can make up a better statistic than the batting average, and many people have (several of these are covered in chapters, and others in the annotated bibliography), but it isn't clear that either the managers or the public are yet ready for them. It is hoped that this book will make both groups more sophisticated so that they will demand better statistics--better in the sense that they are more effective measures of winning potential and therefore more helpful in determining optimal strategies.

Many of the articles in this book are based on work which has previously been published, but published in such widely scattered sources that it is not readily available. In almost every case such articles have been combined, condensed, brought up-to-date, and completely rewritten for this volume. In addition, there are many new articles written especially for this book. Finally, we have diligently searched the literature of the entire world and have summarized, in the annotated bibliography and list of references at the end of the book, everything we think worthwhile which has been published on this subject.

The history of the applications of systems-analytic techniques to sports is very much the history of systems analysis and its applications to many fields of human endeavor. In each such area, the substantive experts always start by saying "your theories may work well in other areas, but in my area of interest everything depends on people, and people are not covered by mathematics". This concept has, of course, been proved false in each area, starting with the military in World War II, when the newly developed field of operations research was responsible for winning the Battle of Britain and the Battle of the Atlantic, and subsequently for innumerable improvements in strategy and tactics in other aspects of the war. Systems analysis is used by hard-headed operators throughout government, industry, and commerce, and increasingly it is being applied successfully to sports.

The first studies of sports were purely descriptive, the earliest such technical articles being described in Pollard's article on Cricket. The first optimization studies were performed in the late 1950's and early 1960's. Many of these were applied to baseball, a particularly suitable topic because the action occurs in discrete events and because the state of the game can be described almost completely merely by specifying the number of outs, the positions of men on base (if any), and the score. Lindsey was among the first to make such studies. His massive article, a complete rewrite and updating of his numerous previous articles on the subject, is the first article in the book, and by far the longest one. While he covers many aspects of the game, his article is supplemented by several others. Palmer, for example, shows a higher-than-expected effect from an additional ball or strike on a batter. Simon analyzes the remarkable recent tendency to seven-game World Series. Freeze, Peterson, and several others indicate that the batting order is not terribly important, and that the order presently adopted is about as good as can be obtained. Cain's article on how the baseball leagues are scheduled is not really on optimal strategies, but we found it fascinating because of the surprisingly large number of constraints which must be faced, and the success with which his computer-aided method meets all these constraints.

While football does not lend itself so easily to analysis, it is also among the most popular American sports, and has the second largest number of articles. There are also articles on several track-and-field events, and on basketball, hockey, golf, rowing, swimming, tennis, and cricket. Finally there are articles which are not confined to a single sport, notably Purdy's impressive summary of the uses of computers in sports. The reader can read the articles in any order, starting with the longest (Lindsey) or the shortest (Palmer), or picking those on his favorite sport. In one or two cases the order is significant; for example, Heffley follows, and improves on, Machol.

We know of no better way to end this introduction than to quote again from Morrison: "Enrico Fermi once remarked that a general was judged great when he had won five or six battles in a row. If battles are fought at nearly even odds, this would happen to about one general in 50 or 100. Among the hundreds of World War II generals, we called several great. This book shows Fermi to be right for World Series teams and managers. For real war the argument is less convincing. Statistics seem to work for baseball; they might work for battles; they certainly do not work nowadays for wars. There the rules keep changing and events are few. World War III will surely not be a doubleheader. Let the analysts rejoice in baseball."[*]

We thank Theresa Bonk for extraordinary dedication to the mechanical aspects of preparing this manuscript.

Shaul P. Ladany and Robert E. Machol

(this ordering of names being a realization of a stochastic process with $\pi = 0.5$).

A WORD TO THE NON-MATHEMATICIAN

by

Robert E. Machol

Few of the articles in this book were written by professional mathematicians (Mosteller's easy-to-read article is an exception), but all of them are written by people who understand and use mathematics effectively. Accordingly, if you have not seen any formal mathematics since high school, you will be at some disadvantage in reading some of these articles. You cannot become a mathematician by reading these few pages, but you can at least learn enough of the technical terms and notational devices to make reading most of these articles comparatively easy. I hope no mathematicians will read the following paragraphs, because I have been a good deal less than rigorous in explaining some technical terms.

Equations, Subscripts, and Summations. A typical equation might be written

$$y = ax \qquad\qquad (1)$$

The number (1) means that this is an equation to which the author will subsequently refer (not all equations are numbered). If he does refer to it he may call it either "equation (1)" or just "(1)". This particular equation is linear (as is the equation $y = ax + b$), because its graph is a straight line (a graph is a sort of picture of an equation). It follows that a particular increment of x always causes the same increment in y. Equations in which there were economies of scale or interactions between variables would be nonlinear, and more difficult to handle. Nonlinear equations include things like $y = x^2$, $y = \log x$, and $z = xy$.

In these equations, x and y represent variables, and a and b represent constants or parameters. Thus, in (1), x might be the length of a golf course in yards, and y the rating (par); a would be a constant because it is always the same for a player of a particular handicap, but x varies from course to course and y varies with it. If we apply the same equation to players with different values for a (different handicaps), then a would be called the parameter (see Scheid's article). That is, a parameter is a number which is constant while considering any one equation, but which might subsequently change.

In addition to equations, mathematicians often use inequations or inequalities, writing the symbols > (greater than), < (less than), $\geq$ (greater than or equal to), and $\leq$ (less than or equal to) in place of = (equals). Thus, $2 < 3$; for another example, $x \leq y$ means x may be less than y or equal to y but cannot be greater than y.

In high school we often used the symbols x, y, and z for "unknowns" and a, b, and c for "knowns". This convention is continued, but we usually refer instead to variables, and to constants or parameters. Furthermore we may need so many of them that we run out of letters. In such cases, many of them are given the same symbol (say x) and are distinguished by a small letter or number written below and to the right, and called a subscript. For example, each observation in a statistical sample is often given the symbol x, with the first observation being called x_1, the second $x_2,...$

(notice the use of three dots to mean "etc."), and the last observation
being x_n, where n is the total number of observations in the sample. If
one then wishes to speak about a generalized observation, without specifying which one, it would be called x_i. An entire set of such numbers is
sometimes called a <u>vector</u>, especially when the numbers are written in the
form of a row or in the form of a column. (The word <u>vector</u> is also used to
apply to a quantity which has both magnitude and direction, and can therefore be represented by an arrow. For example, Brearley distinguishes between
speed, which is simply an undirected rate, and velocity, which is a directed
rate of speed, and is therefore a vector quantity. For technical reasons
it turns out that these two definitions of vector are equivalent.)

    Often it is desired to arrange symbols in both rows and columns. When
this is done, the arrangement is called an <u>array</u> or <u>matrix</u>, and each symbol
has two subscripts affixed to it. For example, in articles by Machol and
by Heffley, a medley swimming relay race is discussed in which there are
four different strokes and many different swimmers. The time of any swimmer in any stroke is given the symbol a; more specifically a_{ki} is the kth
swimmer's time in the ith stroke. Occasionally even more subscripts are
required. For example, Haberman, in discussing football scores, uses the
symbol S_{ijt} to symbolize the score of the ith team against the jth team in
year t. Specifically then, S_{253} might mean the score which Harvard (=2)
achieved against Yale (=5) in the third year of the study. Note that a vector is a special type of matrix with only a single row or a single column.

    It is frequently desirable to add up many such numbers, and the symbol
Σ, which represents the Greek capital letter sigma, is used for this purpose. For example, suppose the x's were the observed values in a statistical sample, so that x_1 was the first observation, x_n was the nth (and last)
observation, and x_i was a typical observation. Then Σx_i would represent the
sum of all the elements in the sample, and we could then define $\bar{x} = \Sigma x_i/n$
to be the sample mean, as explained in the next paragraph. This might be
written more elaborately as $\sum_{i=1}^{n} x_i/n$, or $\sum_{i=1}^{i=n} x_i/n$. As a further example
$\sum_{i=1}^{3} x_i$ would mean $x_1 + x_2 + x_3$; that is, the sum of the first three observations. By appropriate changes of these notational devices, a wide variety
of interesting sums can be displayed. For example, $\sum_{j} S_{2jt}$ would mean the
total score of Harvard against all its all its opponents in year t, and
$\sum_{j} \sum_{t} S_{2jt}$ would be the sum of Harvard's scores against all its opponents in
all the years.

    <u>Probability and Statistics.</u> Every sports fan is familiar with the
term <u>average.</u> Actually, average is a vague word which covers many different things. The most important type of average is called the <u>mean</u>, <u>expec-
tation</u>, or <u>expected value</u>, and is computed by summing all of the observations and dividing by the number of observations. Thus, a baseball player's
batting average (hits/at-bats) or ERA (earned runs/9 x innings pitched) or
the average weight of the defensive linemen on a football team (total
weight/4) would be computed in this way. The mean is usually symbolized
by the Greek letter μ (mu) or by the symbol $\bar{x}$; $\bar{x}$ is used for the mean of a
<u>sample</u>--that is, the mean of a set of observations--while μ is usually used
for the mean of a <u>population</u>, which roughly means the theoretical or assumed

distribution of the random variable. However, if the distribution (of the
random variable) is _skewed_ (asymmetric), the mean value may be unduly affected
by a few extreme values. For example, in determining the handicap of a
golf player, the mean would be inappropriate because most players occasionally have a very high score. A special formula is used there (as explained
in Scheid's article). Often, when the mean is inappropriate, the _median_ or
mode are used. The median is the middle value, the one for which an equal
number fall above and below; the _mode_ is the most common value.

    In addition to the average, one other number is usually sufficient to
describe the nature of a random variable. This number is a measure of the
spread or dispersion, and is called the _standard deviation_. It is defined
in such a way that approximately one third of the occurrences will deviate
from the mean by more than the standard deviation. The standard deviation
is usually referred to by the Greek letter σ (sigma), although the standard deviation of a sample is sometimes designated by the letter s. For
technical reasons, statisticians frequently also talk about the _variance_,
which is the square of the standard deviation (thus if the standard deviation is 3, the variance would be 9), which is given the symbol σ^2.

    The actual nature or shape of the distribution of a random variable--
that is, the way in which small and large values are distributed, and the
probabilities of their occurrences--depends upon the circumstances of the
random process involved. Several of these distribution curves have special
names known to mathematicians, and two of those are important to readers
of this book. One is the _normal_ or _Gaussian_ distribution and is the
familiar bell-shaped curve. For purposes of calculation, this curve is
frequently _standardized_ or _normalized_ so that its average is zero (that is,
half the observations will be positive and the other half negative), and
so that its standard deviation is unity. In this case it is called the
standard normal distribution or the _error function_. Almost every sum or
mean of independent random variables tends to be distributed according to
this formula. Thus, for example, a golfer's total score, which is the
sum of his scores on eighteen individual holes, tends to be normally distributed (as described in Pollock's article). Then if the average were 72
and the standard deviation were four, we would normalize it by subtracting
72 and dividing by 4. Thus, if S were the score, S would be normally distributed, and (S-72)/4 would have a standard normal distribution, which is
easily available in tables.

    The other important formula is the _binomial_ or _Bernoulli_ distribution,
which arises when there is a sequence of _independent_ trials, each of which
has only two possible outcomes (commonly called _success_ and _failure_). This
arises, for example, in baseball, when the possible outcomes of an at-bat
are considered to be only a hit or an out (for other outcomes such as a
walk the trial is not counted). The probability of success on any one
trial is usually given the symbol P or the equivalent Greek letter π (pi);
here we are assuming that this probability does _not_ change from trial to
trial (i.e., all batters are average), which may be an unrealistic assumption. The number of trials is called n; the number of successful outcomes
is frequently called k; and the _proportion_ of successful outcomes (k/n) is
usually called p. Theoretical study of the binomial distribution has
shown that the expected number of successes k is given by $E(k) = n\pi$, and
the standard deviation of the number of successes is given by $\sigma_k = \sqrt{\pi(1-\pi)n}$.
Similarly, the expected proportion of successes is given by $E(p) = \pi$ and

its standard deviation by $\sigma_p = \sqrt{\pi(1-\pi)/n}$. The numerical implications of
these formulas are explored in Lindsey's article.

Regression analysis or least-squares fitting is a statistical technique
used to find out how certain variables affect certain other variables. For
example, Price and Rao use this technique to determine the effect which
such factors as field-goal percentage and number of rebounds have on the
won-loss percentage of a basketball team.

Calculus. A few of the articles in this book use calculus. Most calculus formulas depend on the derivative, which represents a rate of change
and is symbolized by the letter d, often occurring in pairs. For example,
if x is distance and t is time, then dx/dt, called the derivative of x
with respect to t, represents the rate at which distance changes over time,
and is therefore a speed which might be symbolized by v. Then dv/dt would
be the rate of change of speed with time and would be an acceleration.
Such formulas are used in Brearley's article. One nice thing that can be
done with calculus is to find a maximum or minimum (that is, the largest
or smallest value which a function achieves); this is done by setting the
derivative equal to zero and solving the resulting equation.

The other important symbol used in calculus is $\int$, called integral,

and represents a sum, similar to Σ. For example, $x = \int v\,dt$, says that if

you take each tiny increment of time dt and multiply it by the velocity,
v, which occurred during that time, you get the actual distance traversed
during that time, and if you add up all these increments you get the total
distance, x. The simpler formula x = vt would work nicely if v were con-

stant; the advantage of calculus is that $x = \int v\,dt$ allows you to calculate

the distance when v is continuously varying. To symbolize this variation,
the mathematician might write v(t), meaning that v is a function of t
(i.e., the value of v depends on the value of t in accordance with some

equation), and thus write the formula more elaborately: $x = \int v(t)\,dt$. The

same formula might be written still more elaborately: $x = \int_a^b v(t)\,dt$ or

$x = \int_{t=a}^{t=b} v(t)\,dt$, meaning that the integral is to be evaluated (the sum is

to be taken) from the initial time a to the final time b.

Infinity (symbol ∞) is often used by mathematicians to mean "without
limit"; and while it tends to frighten laymen, it usually isn't important.

For example, $x = \int_0^\infty v(t)\,dt$ would mean that the distance traveled from time

zero (an arbitrary starting time) to time infinity is being evaluated. In
this case the object is probably slowing down so that eventually there is
no further significant travel, and the mathematician just doesn't want to
bother with the exact point at which this occurs.

Some Miscellaneous Terms. Stochastic means random; that is, varying
according to some probabilistic rule, as distinguished from deterministic.

Monotonic means continually increasing or continually decreasing, as distinguished from going up and down in alternation. A Markov process is a
system characterized by well-defined states, and by a transition matrix
which gives the probabilities, p_{ij}, that the system will move from state i
to state j. For example, in baseball, the probability that the system will
more from the state {1 out, bases empty} to the state {2 outs, bases empty}
is just the probability that the next batter will be out. A Markov process is distinguished from other stochastic processes by its lack of
memory; the probabilities depend only on the present state, and not on
the history of how that state was reached. Transitivity describes a relationship in which ranking can be consistently ordered. In an intransitive
situation, we might have A beating B, B beating C, and C beating A. Conditional probability is the probability of an event conditional upon specified additional knowledge, and is usually symbolized by a vertical line
separating the symbol for the event from the symbol for the additional know-
ledge; for example, Lindsey writes $p(>0|1,2)$ to symbolize the probability
that there shall be more than zero (i.e. at least one) run, given that there
is one out and a man on second base.

    If all of this seems hopelessly complicated, do not despair. If you
know what an "average" is, and if you know that the average is sometime
called "mean" or "expectation", and if you know that a "standard deviation"
is about how much things usually deviate from the average, you shouldn't
have much trouble.

A SCIENTIFIC APPROACH TO STRATEGY IN BASEBALL

by

G. R. Lindsey

Based on articles in Operations Research, 1959 and 1963,
and Journal of the American Statistical Association, 1961

INTRODUCTION

Baseball should be particularly well suited to scientific analysis.
Unlike soccer, ice-hockey, basketball, and other continuously moving sports,
it is a stop-and-go game, with frequent pauses at which the state of play
can be specified by a small number of variables, and decisions taken regarding the strategy for the next play. Moreover, certain types of statistics, the subject of the avid attention of millions of followers, have been
kept in a uniform manner for a period approaching a century [Turkin and
Thompson, 1956; Neft et al, 1974], during three quarters of which the rules
of the game having remained substantially unchanged. Those changes which
have occurred, such as the increased resilience of the ball, the prohibi-
tion of trick deliveries such as the spitball, or the introduction of the
designated hitter, have followed long intervals of uniform rules, which
should invite statistical analysis of the effects of the changes.

There is no dearth of stored data regarding individual and team accom-
plishments in the past. Broadcasters keep up a continual barrage of statistical information bearing on the past performance of the players in the
game in progress, and if anything unusual should occur, they are usually
able to tell how many times it has happened before.

The gap in the analysis of baseball which appears evident to a prac-
titioner of operational research or management science is any theory that
could be used to support the strategic decisions that must be made during
the game. In this context we will exclude decisions made while the ball
is in play.

LIMITATIONS TO THE VALUE OF INDIVIDUAL RECORDS

Past Performance as a Predictor of Future Performance. If a batter
has made 25 hits in 100 official Times at Bat since the beginning of the
season, his batting average of .250 can be regarded, quite accurately, as
a measure of his average past performance so far this season. But if this
same batter has, over the past several years, accumulated 1500 hits in
5000 Times at Bat, then his "Lifetime Average" is .300. This is also an
accurate measure of past performance, now averaged over several seasons.

What does this tell us about his performance next time he comes to
bat? In the next game? For the rest of this season? Can we adopt the
hypothesis that he has a "real" batting average, constant through his major
league career both past and future? If so, our best measure of this real
average is .300, which can be restated as a probability of 0.300 that he
will make a hit for each Time at Bat. His performance so far this season
does not invalidate the hypothesis, since the probability of deviating by
5 or more from the 30 expected successes in 100 independent trials in each

of which the probability of success is 0.3 is about 28%. But if his average
for this season were still .250 after 500 Times at Bat, it would now be
highly questionable that the "true probability of a hit" had been .300
throughout this season. A fluctuation down to 125 or fewer successes in
500 trials would occur with a probability below 1%. Thus it takes a certain amount of history before an inference can be drawn, and the higher
the level of confidence demanded, the more history is required.

But what about the basic hypothesis -- that a batter has a constant
probability of making a hit? May his probability of making a hit not depend on a host of factors such as whether he is facing a right- or left-
hand pitcher, a very fast pitcher or a curve- or knuckle-ball specialist,
whether the game is being played at home or away, or by day or by night?
May the batter usually do better in certain parks, perhaps because of the
layout of the fences? Or is it not possible that his physical prowess is
deteriorating with age, and his real current average is less than in previous years?

The Example of Right- and Left-Handedness. It has always been believed
that a right-handed batter does better against a left-handed pitcher than
against a right-handed pitcher. Similarly, it is believed that a left-
handed batter does better when the pitcher has the opposite "handedness".
This hypothesis was verified by analysis of over 12000 Times at Bat in the
National, American, and International Leagues for the 1951 and 1952 seasons.
The data was collected by the author and by Lieut. Colonel Charles Lindsey,
by following the progress of games seen directly, on television, or listened
to on the radio. Newspaper box scores were not used, and batters known to
be ambidextrous "Switch Hitters" were excluded. The results are shown on
Table I.

The last column gives $\sigma = \sqrt{p(1-p)/N}$, the standard deviation of the
mean of a sample from a binomial distribution, where $p = H/(AB)$ = the
calculated Batting Average, N is the number of Times at Bat (AB), and H
is the number of hits.

Examination of Table I suggests that there is no significant difference between the two categories of "Same", the average being .231, and no
significant difference between the two categories of "opposite", the average being .263. But the difference between "Same" and "Opposite", .032,
with a standard deviation of about .008, is highly significant, the probability of a difference of 4σ being about 0.00006. Thus it can be safely
concluded that, averaged over all the circumstances of hundreds of games,
there is indeed an advantage to the batter if the pitcher is of the opposite hand.

Suppose then that a manager is prepared to accept the proposition
that an average batter has a batting average better by .032 against a
pitcher of the opposite hand than of the same "handedness". Suppose that
R, a right-hand batter, is due to face a right-handed pitcher P in a crucial situation when a hit is badly needed. The manager of the batting team
is considering the advisability of replacing R by a left-hand pinch-hitter
L. If R, L, and P were all "average" performers, the criterion could be
that L's batting average be no farther than .032 below that of R.

However, before he accepts such a simple criterion, the manager should
ask himself a number of questions. Are the players in question "average"?

Table I

Averages of Right- and Left-Handed Batters
Against Right- and Left-Handed Pitchers

Batter	Pitcher	AB	H	B.Avg.	σ
R	R	5197	1201	.231	.006
L	L	1164	270	.232	.012
Same		6361	1471	.231	.0053
L	R	4002	1055	.264	.007
R	L	2245	590	.263	.009
Opposite		6247	1645	.263	.0056
All		12608	3116	.247	

May R be less (or more) susceptible than most right-handed batters to the tricks of right-handed pitchers? Is P particularly effective against batters of the same handedness? Has R (or L) been especially successful (or unsuccessful) against this particular pitcher P? Does the fact that the game is being played at night (in the daytime, at home, on the road) affect the probability of a hit being made by R (or L)?

The Problem of Sampling Error. If the manager wishes to have answers to these questions from past statistics, he is very likely to be frustrated by sampling errors. It may indeed be possible to record the batting averages of his individual players against right- and left-handed pitchers, and to accumulate sufficient data to establish the averages within a reasonably useful margin of error. After all, pitchers have only two arms. If the batting average of L against all the right-handed pitchers he has faced exceeds that of R, it would seem advisable to have L bat. But as we have seen, this difference may not be statistically significant unless it has been possible to record the performance of both R and L over several hundred Times at Bat. More valuable information would be the averages of R and L while batting against P. But no batter is likely to face any one pitcher more often than about twenty times in an entire season--usually it will be far less often than that. Normally, individual one-against-one data will be too sparse to yield significant results. It would be possible to obtain averages with reasonably small sampling errors for all games home and away (up to eighty for each per season) or by day and by night (with day games less frequent, except for the Chicago Cubs). But can one separate the effects of handedness, home and away, day or night, etc., perhaps by using analysis of variance? Once these categories are subdivided the numbers for each combination will likely be too small to reveal small differences in performance significant at a meaningful level of confidence.

It may be that an experienced manager is able to generalize from acute observation of playing styles. Based on personal observation of many thousands of Times at Bat, he learns which batting styles are most likely to succeed against different styles of pitching. He may also learn more from a player's demeanour than from his recent statistics, and he may wish to make allowances for recent runs of good or bad fortune which may have brought "easy" hits or "robberies".

THE PROBABILITY OF WINNING THE GAME

One could select various objectives for baseball strategy. In the short range it could be to get on base, to get a man out, to score a run, or to prevent a run being scored. More distant goals could be to maximize (or minimize) the number of runs scored in an inning, or to win the game. Very long-range objectives would be to win the championship, or to maximize the profit at the gate. These are not all entirely consistent, as will be shown later. To cite an elementary example, the objective of a team batting in the last half of the ninth inning with the score tied is surely to maximize the probability of scoring at least one run, while for a team batting in the first half of the first inning, the proper objective is presumably to maximize the expected number of runs (and this conclusion is quantitatively justified below).

We will take as our objective the winning of the game in progress. This will almost always be consistent with longer-range objectives, although one could imagine a situation in which the use of an injured or exhausted player in order to increase the probability of winning today's game could risk aggravation of his problem, with consequent loss of subsequent games and ultimately of the championship.

Let us define the situation at any moment between plays by five variables.

ℓ, the lead in runs, i.e., the difference in the number of runs that have been scored by the team whose strategy is under consideration and the other team. $\ell = 0, \pm 1, \pm 2, \ldots$

i, the inning being played (or about to be started) $i = 1, 2, \ldots$

a binary variable identifying which half of the inning is being played (or about to be started)

- V if the Visiting Team is at bat, H if the Home Team is batting.

T, the number of men already out in the current half-inning $T = 0, 1, 2$.

B, the situation on the bases. The eight possible states of occupation are indicated by $B = 0, 1, 2, 3, 12, 13, 23, F$.

This formulation assumes that the situation within the half-inning can be completely described by the two variables $\{T, B\}$, which occur in twenty-four combinations. It neglects the ball-and-strike count, a questionable simplification of some significance (see article by Palmer, page 31).

We now define a probability distribution $P(r|T,B)$, giving the probability that, between the time that a batter comes to the plate with T men out and the bases in state B, and the end of the half-inning, the team will score exactly r runs. This function $P(r|T,B)$ allows us to convert the present situation within the half-inning to the end of the half-inning. Thus if the Home Team have a lead of ℓ runs (which may be positive, negative, or zero), and are batting in the ith inning with T men out and the bases in state B, then the probability that their lead will be $(\ell+r)$ at the end of half-inning Hi is $P(r|T,B)$.

We now need to determine the probability that, if they complete half-inning Hi with a lead of $(\ell+r)$, they will eventually win the game. This requires the other major probability distribution function. Let us define $_{Hi}W(\ell)$ as the probability that, if the Home Team has a lead of ℓ runs at the end of their half of the ith inning, they will eventually win the game.

These two distributions can now be combined to compute $\Omega(H,i,\ell,T,B)$ the probability that the Home Team will win the game, if they have a lead of ℓ runs and are batting in the ith inning with T men out and the situation on the bases is B.

$$\Omega(H,i,\ell,T,B) = \sum_{r=0}^{\infty} P(r|T,B) \,_{Hi}W(\ell+r),$$

which sums the probabilities of winning should H score 0, 1, 2, ... runs before the end of the ith inning. Similarly, when the Visiting Team is batting, we have

$$\Omega(V,i,\ell,T,B) = \sum_{r=0}^{\infty} P(r|T,B) \,_{Vi}W(\ell+r)$$

THE BUILDUP AND OVERCOMING OF A LEAD

It now remains to derive these two distribution functions. We begin with $W(\ell)$. A method of obtaining $W(\ell)$ by direct statistical estimation would be to record, from a large number of games, the proportion of times that the Home Team won after it had established a lead of ℓ at the end of its half of the inning. Data was collected from the inning-by-inning linescores of the last 782 games played in the National, American, and International League season of 1958. Table II shows the information needed for direct estimation of $_{Hi}W(\ell)$. The three rows of numbers for each inning show the number of times that a lead of ℓ was held at the end of Hi by the team that eventually won the game, the number of times the same lead ℓ was held by the team that lost the game, and the probability $_{Hi}W(-\ell)$, calculated from those two numbers, that a team trailing by ℓ runs at the end of Hi will overcome the deficit and eventually win the game. The probability that a team leading by ℓ at the end of Hi will win is obviously

$$_{Hi}W(\ell) = 1 - \,_{Hi}W(-\ell)$$

Table II

Frequency of Establishing and Overcoming of a Lead
of ℓ in the 782 games of 1958

Inning	lead ℓ :	1	2	3	4	5	≥ 6
1	W	133	74	27	17	6	6
	L	59	31	4	1	3	1
	$_{H1}W(-\ell)$	.31	.30	.13	.06		
2	W	175	105	40	24	14	13
	L	79	30	10	4	2	0
	$_{H2}W(-\ell)$	.31	.22	.20	.14	.1	.0
3	W	152	99	71	51	29	36
	L	87	32	12	8	2	0
	$_{H3}W(-\ell)$	.36	.24	.14	.14	.06	.00
4	W	167	96	94	59	37	44
	L	80	42	11	3	2	0
	$_{H4}W(-\ell)$	.32	.30	.10	.05	.05	.00
5	W	139	128	80	73	43	68
	L	72	39	7	3	1	1
	$_{H5}W(-\ell)$	.34	.23	.08	.04	.02	.01
6	W	134	128	94	81	50	84
	L	53	26	6	4	1	0
	$_{H6}W(-\ell)$	.28	.17	.06	.05	.02	.00
7	W	128	129	109	77	58	112
	L	36	15	4	2	0	0
	$_{H7}W(-\ell)$	.22	.10	.04	.03	.00	.00
8	W	141	136	101	89	61	139
	L	19	10	0	1	0	0
	$_{H8}W(-\ell)$	.12	.07	.00	.01	.00	.00

For example, in 192 of the 782 games, a lead of 1 run had been established at the end of the first inning. In 133 of these games the team that
was 1 run ahead at the end of H1 eventually won the game; in the other 59
games it eventually lost. So we estimate the probability that the game
will be won by a team trailing by one at the end of H1 to be $_{H1}W(-1) =$
59/192 = 0.307, with a standard deviation of $\sqrt{(.307)(.693)/192}$ = .033.

It can be seen from Table II that a lead of more than $\ell=2$ is not
attained very often in the early innings, or overcome very often in the
late innings, so that the sampling error of the estimated $_{Hi}W(-\ell)$ can be
quite high. For example, we would infer from Table II that a team which
trailed by five runs after one inning had one chance in three of ultimately
winning the game; the correct probability (Table IV) is less than one chance
in ten. To obtain a full set of values of $_{Hi}W(-\ell)$ with a low probable error
by this direct statistical method, it would be necessary to collect the results of many thousands of games.

   When faced with this type of difficulty, an alternative approach is
to construct a mathematical model of the process under study--in this case
the buildup of the score as the innings progress. If the model can be
validated by the statistics of real games, it may yield the desired information without the need for an enormous sample.

   Distribution of the Scores by Half-Innings. Analysis of the data from
the 782 games of 1958 showed that there were differences between the distributions of runs scored in different innings, statistically significant well
beyond the 1% level of confidence. In particular, the mean scores were
greatest in the first and third innings, and much smaller in the second
inning (presumably because the batting order is designed to produce maximum effectiveness in the first inning; the weak tail of the order tends to
come up in the second inning, the strong head to reappear in the third).
It is evidently not valid to assume identical distributions of scoring for
each inning.

   Table III shows further data of the same nature taken from 1000 games
selected at random from the National and American Leagues in 1959. Games
abandoned due to weather were excluded. The columns headed 0 to > 5 show
the relative frequency with which one team scored r runs in the ith inning.
N_i shows the number of ith half-innings recorded. The third-from-last
column shows the mean numbers of runs scored by one team in the ith inning,
the penultimate column gives the standard deviation of the number of runs
scored, and the last column the standard deviation of the mean. The rows
after the eighth inning are labelled as follows:

   C Completed half-innings ($1 \leq i \leq 8$ are all completed)

   I Incompleted half-innings (because the game was won
      and finished before three men were out)

   E Extra innings (i=10, 11, ...)

   It is immediately evident from the last three columns that many of
these distributions are significantly different from one another.

   The top half of Figure I shows the mean number of runs per half-inning
for both samples, indicated by vertical bars of length $2\sigma/\sqrt{N_i}$ (i.e., $\pm$ one
standard deviation of the mean). The solid black bars are for the 782
games of 1958, the hollow white bars for the 1000 games of 1959. In the
bottom half, the "big innings" are indicated by showing the probability of
scoring 3 or more runs. There is no significant difference between the
sets of data for the two seasons.

   The distributions in Table III are very different for complete and
incomplete innings, because the incomplete half-innings are necessarily
the half-innings in which the games were won (else they would not be incomplete). They must therefore have at least one run, and it can be seen
from Table III that many have two or three runs. The mean score for all
incomplete half-innings is 1.66, as contrasted to only 0.487 for all completed half-innings. When this higher-than-average sample is extracted,
the remaining residue (of completed half-innings) is below-average, with
mean scores 0.40 for 9C and 0.32 for EC.

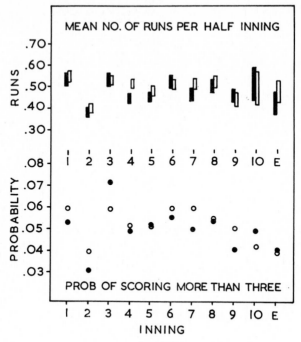

Figure 1
Scoring by Individual Half-innings
Black: 782 games of 1958
White: 1000 games of 1959

Since we have determined that there is a significant difference in the distribution of runs for different innings, our model of the buildup of the score must be made from separate distributions $f_i(r)$, one for each inning. Because of the comparatively small number of extra innings, the large sampling error for individual extra innings suggests that we lump all completed extra half-innings together into a composite distribution $f_E(r)$. The model game now consists of two random drawings from each distribution $f_1(r)$ to $f_9(r)$, one for the Visiting Team V and one for the Home Team H. If the score is tied at the end of H9, repeated pairs of drawings are made out of the extra-inning distribution $f_E(r)$ until the game is won. As a first-order refinement, a win in H9 or HE by more than one run would be reduced to one run, assuming that the game ceased as soon as H got ahead by one in the ninth or any extra inning.

This model has an inherent assumption that the distribution of runs scored subsequent to the completion of any half-inning is unaffected by the scoring history of the game previous to that time. There are a priori reasons to doubt this assumption of independence. Pitchers may tire as runs begin to accumulate. Once a lead has been established, the strategies may alter as the leading team tries to maintain its lead and the trailing team tries to overcome it. But if the assumption of independence is valid, then it is not difficult to calculate $W(\ell)$, the probability of winning given a lead of ℓ runs.

Inning (i)	r = Number of Runs Scored							N_i	Mean	σ	$\sigma/\sqrt{N_i}$
	0	1	2	3	4	5	>5				
1	.700	.159	.081	.031	.018	.010	.001	2,000	.54	1.03	.02
2	.768	.139	.053	.023	.009	.003	.005	2,000	.40	.92	.02
3	.730	.131	.079	.029	.016	.009	.006	2,000	.53	1.10	.03
4	.719	.151	.078	.027	.013	.007	.005	2,000	.51	1.03	.02
5	.730	.145	.074	.033	.011	.005	.002	2,000	.48	.98	.02
6	.721	.157	.062	.036	.012	.008	.004	2,000	.51	1.03	.02
7	.731	.138	.071	.029	.020	.008	.003	2,000	.51	1.04	.02
8	.710	.162	.073	.027	.019	.006	.003	2,000	.52	1.03	.02
9C	.770	.129	.060	.027	.010	.004	.000	1,576	.40	.87	.02
9I	.000	.46	.26	.23	.03	.02	.00	61	1.88	.98	.13
EC	.825	.112	.033	.012	.003	.006	.009	331	.32	.95	.05
EI	.00	.73	.17	.06	.04	.00	.00	53	1.40	.76	.10
All E	.711	.198	.052	.018	.008	.005	.008	384	.47	1.00	.05
All C	.730	.146	.070	.029	.014	.007	.004	17,907	.487	1.01	.008
All I	.00	.59	.22	.15	.03	.01	.00	114	1.66	.92	.09
All	.726	.148	.071	.030	.014	.007	.004	18,021	.493	1.05	.008

Table III

The Relative Frequency with Which One Team Scored r Runs in the ith Inning

(Based on 1000 Games from NL, AL 1959)

The basic function needed for calculating $W(\ell)$ is $f_i(r)$, the probability that a team will score r runs in its half of the ith inning, provided that the half-inning is played, and assuming it to be played to completion. If half-innings were always played to completion, the distribution observed in Table III could provide direct estimates of $f_i(r)$. However, for $i \geq 9$ the distributions for completed half-innings recorded on
the table cannot be used directly, since they have excluded the decisive
winning home half-innings, for which $r > 0$, and consequently are a biased
sample, while the distributions for incomplete half-innings are ineligible
just because they are incomplete.

By assuming that all games won in incomplete half-innings are won by
a margin of one run, calculations outlined in Appendix B of [Lindsey, 1961]
allow us to deduce $f_9(r)$ and $f_E(r)$ from the measured distributions for 9C,
9I, EC, and EI. The resulting $f_9(r)$ is very close to the mean distribution
obtained by averaging $f_i(r)$ for $1 \leq i \leq 8$. If we assume this mean distributions for 9C and 9I, the agreement with Table III is excellent for the
1576 completed half-innings 9C. It is less satisfactory for the 61 incompleted half-innings 9I, but the discrepancy may be due to the original
assumption that the margin for a game won in H9 is always one run. In
fact, it could be as high as four should a grand-slam home run be hit.
Since the discrepancy is only significant at the 4% confidence level, only
for the incomplete half-innings, and probably explained by the inaccuracy
of the assumption of the one-run win, we conclude that there is nothing
unusual about the scoring pattern in the ninth inning, except the predictable
effect of truncating the game as soon as H establishes a lead. So for further calculation, $f_9(r)$ is taken as the mean of the distributions for the
first eight innings.

If we assume that $f_E(r)$ has the same mean distribution as $f_9(r)$, and
predict the distributions for EC and EI, there is good agreement for the
331 completed extra innings EC of Table III, but poor agreement for the 53
incomplete extra innings EI. Again, fewer one-run cases are observed than
predicted.

Now that we have our distributions $f_i(r)$, we would like to determine
whether they are independent from half-inning to half-inning. Some statistical tests, described in Appendix C of [Lindsey, 1961], using correlation
coefficients and conditional distributions, revealed no evidence of dependence between half-innings. As a final step in the validation of the model,
now assumed to be inhomogeneous but independent as between half-innings,
several distributions were calculated which could be directly compared with
the results actually experienced in the 1958 and 1959 seasons.

The Length of a Game, the Total Score, and the Buildup of a Lead. If
we know $f_i(r)$, the probability that either team will score exactly r runs
in the ith inning (if it is played to completion), we can compute the probability that games will require extra innings. For example, as demonstrated
in [Lindsey, 1961], it is possible to calculate $G_{ij}(x)$, the probability that
one team will increase their lead by exactly x runs between the beginning
of the ith and the end of the jth inning. The probability that a team will
have established a lead of exactly ℓ runs by the end of the ith inning is
$G_{1i}(\ell)$, $1 \leq i \leq 8$. In particular, the probability that a tenth inning will
be required is $G_{19}(0) = 0.106$, and the probability that an extra inning
will not produce a decision is $G_{EE}(0) = 0.566$. The experience in 1777
major league games played in 1958 and 1959 agreed with these predictions
at a significance level of 10%.

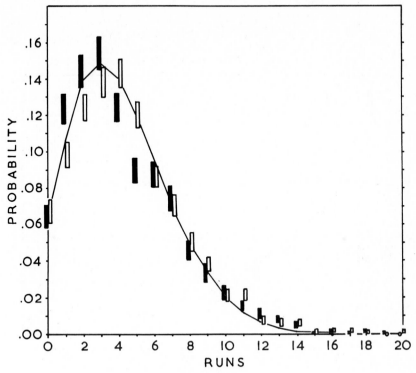

Figure 2
Distribution of Total Runs
Scored by One Team in a Game
Black: 782 games of 1958
White: 1000 games of 1959

The total score of one team in a game can also be calculated, as shown
in Figure 2. The midpoints of the vertical rectangles represent the distribution of scores actually observed, from the 782 games of 1958 in black
and the 1000 games of 1959 in white. The lengths of the bars show $\pm$ one
standard deviation. About two-thirds of the bars intersect the line, which
would be expected if the fit were perfect.

Calculation of the final winning margin requires treatment of the
cases that H9 is not needed, and that the game is won in H9 or HE with the
decisive half-inning being incomplete. The expressions are derived in
Appendix B of [Lindsey, 1961]. The actual experience in the 1958 and 1959
seasons conforms very well to the predicted distribution. The final winning
margins are shown on Figure 3, on which the continuous line shows the margin calculated from the model, while the vertical black bars show the results observed for the 782 games of 1958 and the vertical white bars for
the 1000 games of 1959. The center of the bar marks the observed frequency,
and the total length of the bar is $2\sqrt{p(1-p)/N}$, where the predicted frequency is p. The fit is very satisfactory.

Holding or Overcoming a Lead, and Winning. Define the probability
that a team will increase its lead by ℓ runs or more, between the beginning
of the ith and end of the jth innings, as

Figure 3
Distribution of Final Winning Margins
Black: 782 games of 1958
White: 1000 games of 1959

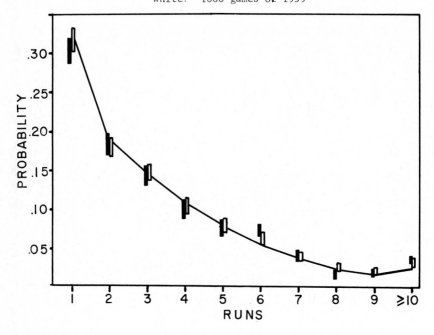

$$U_{ij}(\ell) = \sum_{k=\ell}^{\infty} G_{ij}(r) \qquad \text{for} \qquad i \leq j \leq 9, \ -\infty < \ell < \infty$$

(G is the probability of increasing lead by exactly ℓ,
U by ℓ or more).

If Team A is exactly ℓ runs ahead at the end of Hi, the probability that
their lead will remain at 1 or more at the end of H9, which is the probability that they will win the game in 9 innings, is $U_{i+1,9}(1-\ell)$. The probability that their lead will be annulled and the game require extra innings is $G_{i+1,9}(-\ell)$. We assume that in this event the probability of A
winning is $\frac{1}{2}$.

At last we have our formula for the probability that A will win the
game, if they have a lead of ℓ at the end of Hi. It is

$$_{Hi}W(\ell) = U_{i+1,9}(1-\ell) + \tfrac{1}{2} G_{i+1,9}(-\ell)$$

$_{Hi}W(\ell) + _{Hi}W(-\ell) = 1$, so that the function is symmetrical about $\ell=0$,
and the probability can be applied to either team.

Finally, we will also need the probability that the Visiting Team V will win the game if they have a lead of ℓ at the end of their half of the ith inning. This can be calculated from the relation

$$_{Vi}W(\ell) = \sum_{r=0}^{\infty} f_i(r)\ _{Hi}W(\ell-r) \qquad \text{for } i = 1,2,\ldots 8$$

$$\text{and } _{V9}W(\ell) = \sum_{r=0}^{\ell-1} f_9(r) + \tfrac{1}{2} f_9(\ell).$$

$_{Vi}W(\ell)$ is not symmetrical about $\ell=0$, since, at the end of Vi, the Home Team have one more half-inning than the Visitors in which to score. The probability that the Home Team H will win if they have a lead of ℓ at the end of Vi is $1 - _{Vi}W(-\ell)$.

The probabilities $_{Hi}W(\ell)$ and $_{Vi}W(\ell)$ calculated from the model are tabulated on Table IV, and shown in slightly smoothed graphical form on Figures 4 and 5. If useful for no other purpose, this information should be invaluable for those disposed to make wagers in the course of games as to their final outcomes. $_{V1}W(\ell)$ can be calculated for negative values of ℓ, but these numbers (shown in brackets on Table IV) have no significance in an ordinary baseball game, since the Visiting Team cannot be behind at the end of their half of the first inning. However, as an illustration of the ability of a model to extend prediction beyond the area measurable from past experience, these figures could be used in a tournament using a handicap, or a two-game series based on total scores. In this case V could begin half-inning V1 with a deficit.

It is evident from both Table IV and Figures 4 and 5 that the probability of overcoming a given deficit decreases as the game proceeds. This can be interpreted as saying that the value (for winning the game) of a lead of a given number of runs increases as the game progresses. Moreover, the value of one additional run, $W(\ell+1) - W(\ell)$, is greatest for $\ell=0$, successively smaller for $\ell=-1$ and for $\ell=1$, and continues to decrease for larger absolute values of ℓ. In other words, the most valuable run is the "gamer" which ties the score or puts the team one run ahead. And as the game progresses, the tying and lead runs ($\ell=0,1$) become relatively more and more important. For example, changing ℓ from 0 to 1 increases the probability that V will eventually win the game by 0.105 in the first inning, but by 0.440 in the ninth inning (and, of course, by 0.5 for H in the ninth inning).

This fact is of considerable importance in setting objectives for field strategy. In the early innings, for which the contours of constant ℓ are nearly equally spaced, the value of 1, 2, 3, ... runs is nearly proportional to their number. To maximize the expected number of runs in an early half-inning is almost the same as maximizing the probability of winning the game. But in late innings the value of runs for increasing the probability of winning is far from a linear function of the number of runs. The paramount objective becomes tying or getting a lead of one run.

Table IV

Calculated Probability That Visiting Team With Lead of ℓ Runs
at End of Half-Inning i will Win the Game

$$_{Vi}W(\ell)$$

Lead	V1	V2	V3	V4	V5	V6	V7	V8	V9/VE
6	.917	.928	.935	.953	.966	.973	.984	.993	.997
5	.873	.891	.899	.922	.940	.949	.970	.980	.992
4	.818	.839	.844	.873	.896	.912	.940	.960	.983
3	.744	.767	.772	.801	.828	.889	.889	.924	.962
2	.655	.680	.680	.708	.730	.752	.800	.848	.913
1	.553	.574	.566	.585	.598	.610	.652	.702	.808
0	.448	.457	.442	.443	.437	.420	.416	.390	.363
-1	(.343)	.348	.321	.309	.287	.246	.203	.122	0
-2	(.253)	.251	.221	.204	.177	.139	.101	.050	0
-3	(.178)	.175	.145	.129	.104	.076	.050	.020	0
-4	(.121)	.118	.092	.078	.058	.041	.024	.009	0
-5	(.080)	.076	.056	.045	.032	.020	.011	.003	0
-6	(.051)	.048	.033	.024	.016	.009	.004	.000	0

$$_{Hi}W(\ell)$$

Lead	H1	H2	H3	H4	H5	H6	H7	H8	H9/HE
6	.939	.944	.959	.970	.979	.988	.994	.998	1
5	.905	.913	.932	.947	.961	.974	.987	.995	1
4	.857	.867	.889	.908	.930	.949	.970	.987	1
3	.793	.803	.827	.849	.877	.907	.940	.974	1
2	.710	.720	.741	.764	.793	.831	.878	.936	1
1	.610	.617	.630	.647	.668	.705	.756	.846	1
0	.500	.500	.500	.500	.500	.500	.500	.500	.500
-1	.389	.383	.368	.353	.331	.295	.244	.153	0
-2	.290	.280	.257	.236	.207	.168	.122	.063	0
-3	.207	.196	.171	.150	.122	.093	.060	.025	0
-4	.142	.133	.109	.091	.070	.050	.029	.011	0
-5	.095	.087	.067	.053	.038	.025	.013	.004	0
-6	.061	.055	.039	.029	.020	.012	.005	.000	0

Finally, it remains to compare our experimentally observed values of $_{Hi}W(\ell)$, listed in Table II, with the prediction based on the model. Figure 6A shows the (unsmoothed) predictions of Table IV for odd deficits $\ell = -1$, -3, -5, and -7 as the continuous lines, and Fig. 6B for even deficits $\ell = -2$, -4, -6, and -8. The observed results, shown as small circles, represent the proportion of games in which one team experienced the stated deficit $(-\ell)$ at the end of Hi, but subsequently won. The vertical bars show the standard deviation of the estimate, as calculated from the data of Table II.

Table V

Distribution of Scores in Remainder of Half-Inning

Data	B	T	N(T,B)	P(0|T,B)	P(1|T,B)	P(2|T,B)	P(>2|T,B)	E(T,B)	$\sigma/\sqrt{N}$
59/60	0	0	6561	.747	.136	.068	.049	.461	.012
	0	1	4664	.855	.085	.039	.021	.243	.011
	0	2	3710	.933	.042	.018	.007	.102	.008
59/60	1	0	1728	.604	.166	.127	.103	.813	.031
	1	1	2063	.734	.124	.092	.050	.498	.022
	1	2	2119	.886	.045	.048	.021	.219	.016
59/60	2	0	294	.381	.344	.129	.146	1.194	.083
	2	1	657	.610	.224	.104	.062	.671	.043
	2	2	779	.788	.158	.038	.016	.297	.024
59/60	3	0	67	.12	.64	.11	.13	1.39	.09
	3	1	202	.307	.529	.104	.060	.980	.072
	3	2	327	.738	.208	.030	.024	.355	.040
59/60	12	0	367	.395	.220	.131	.254	1.471	.087
	12	1	700	.571	.163	.119	.147	.939	.051
	12	2	896	.791	.100	.061	.048	.043	.032
59/60	13	0	119	.13	.41	.18	.28	1.94	.15
	13	1	305	.367	.400	.105	.128	1.115	.077
	13	2	419	.717	.167	.045	.071	.532	.054
59/60	23	0	73	.18	.25	.26	.31	1.96	.18
	23	1	176	.27	.24	.28	.21	1.56	.10
	23	2	211	.668	.095	.170	.067	.687	.080
59/60	F	0	92	.18	.26	.21	.35	2.22	.20
	F	1	215	.303	.242	.172	.283	1.642	.105
	F	2	283	.671	.092	.102	.135	.823	.085
		$\Sigma N = 27027$							
52/60	F	0	173	.17	.27	.17	.39	2.254	.145
	F	1	419	.310	.242	.186	.262	1.632	.080
	F	2	527	.645	.114	.110	.131	.861	.06

To illustrate, Table IV predicts that a team two runs behind at the end of the fourth inning has a probability of 0.236 of winning the game. Table II shows that a lead of two ($|\ell| = 2$) was established in 96+42=138 games (out of 782), out of which the team trailing by two at the end of H4 eventually won 42. Hence the estimated value of $_{H4}W(-2)$ is 42/138 = .303. The difference (Observed - Expected) of .067 is about 1.8 times the standard deviation $\sigma = \sqrt{(.236)(.764)/138} = .037$, indicated by the length of the vertical bar. There is a probability of about 7% of obtaining a de-

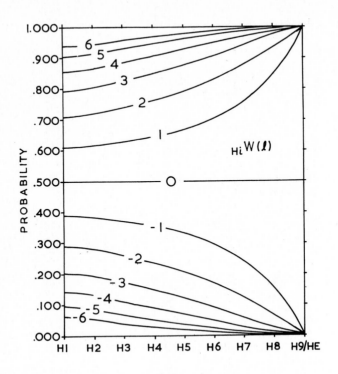

Figure 4
The Probability that a Team Will Win the Game, if they are ℓ
Runs Ahead at the End of the Home Half of the ith Inning

viation as great or greater than $\pm 1.8\sigma$, so that the observation does not demand rejection of the model. The agreement between the calculated curves
and the observed points for $\ell = -1$ and $\ell = -3$ on Fig. 6A, and $\ell = -2$ and $\ell = -4$
on Fig. 6B is quite satisfactory, taking account of the substantial sampling
errors indicated by the lengths of the vertical bars, and of the fact that
about 32% of them would "miss" the calculated points if the model were perfect. It would seem entirely justified to adopt the values of $W(\ell)$ in
Table IV, calculated from the model of scoring, as a valid basis for further
deductions.

THE SCORING IN THE REMAINDER OF A HALF-INNING

With $W(\ell)$ calculated, it remains to determine $P(r|T,B)$, the probabil-
ity that between the time that a batter comes to the plate with T men out
and the bases in state B, and the end of the half-inning, the team will
score exactly r runs.

To determine these distributions from actual baseball experience, a
large number of games were observed (by radio, television, or personal
attendance) and data were recorded in such a way that the number of runs

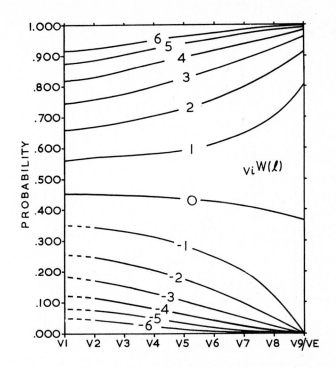

Figure 5
The Probability that the Visiting Team Will Win the Game, if They
are ℓ Runs Ahead at the End of Their Half of the ith Inning

scored subsequent to the occurrence of each new 'situation' as defined above
could be tabulated. If the situation changed while the batter was still at
the plate (e.g., by a stolen base, wild pitch, or runner being picked off)
only the situation pertaining when he first came up was recorded. If a
half-inning was not completed (e.g., in a game won in the last half of the
ninth or an extra inning), nothing was recorded. Data taken from all completed half-innings was pooled.

Exclusion of incomplete half-innings will introduce a bias into the
data, since, with the rare exception of games abandoned because of weather,
a half-inning is incomplete because the winning run has been scored, and
the average score in an incomplete half-inning is greater than in a complete half-inning. However, the proportion of all half-innings that are
incomplete is less than 0.7 per cent, and the proportion of runs scored in
incomplete half-innings is less than 2.2 per cent, so that the bias is small.

The majority of the recording and tabulation was contributed by LCol.
Charles Lindsey. Most of it came from 3033 half-innings in 1959 (from all
or parts of 176 major-league games) and 3366 half-innings in 1960 (from
all or parts of 197 major-league games). For some purposes use was also

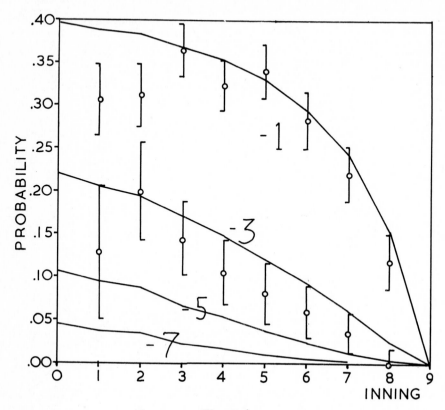

Figure 6a
Probability of Overcoming a Lead
and Winning 782 Games of 1958
(Odd leads only)

made of data taken in 1952, 56, 57, and 58 concerning only situations with
the bases full. A summary of the results is given in Table V.

The column headed N(T,B) gives the number of cases recorded. It will
be seen that there is a very wide variation. Over half of the 27,027 cases
are for the bases empty. Some situations ($\{0,3\},\{0,23\},\{0,F\}$) each occur
only once in about 300 cases. In addition to values of $P(r|T,B)$, the prob-
ability that r runs will be scored, Table V also shows the expected number
of runs.

$$E(T,B) = \sum_{r=0}^{\infty} rP(r|T,B)$$

and $\sigma/\sqrt{N}$, the standard deviation of $E(T,B)$, as computed from these statistics. As would be expected, the values of $E(T,B)$ show a sharp reduction
as T rises from 0 to 1 to 2 out.

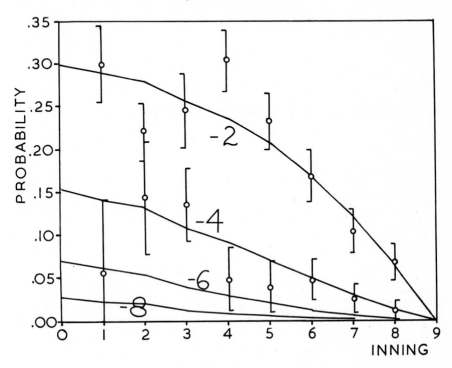

Figure 6b
Probability of Overcoming a Lead
and Winning 782 Games of 1958
(Even leads only)

DECISION RULES FOR CERTAIN SITUATIONS

The Intentional Base on Balls. For most of the strategies to be discussed, it is never certain that the intended play will or even can be carried out as intended. However, the issuing of an intentional base on balls
is completely under the control of the fielding side, barring such improbable incidents as a wild pitch or an attempt by the batter to hit.

There are two usual purposes behind the decision to issue an intentional walk. One is to avoid pitching to a particularly dangerous batter
who is followed in the order by a much weaker batter. The other is in the
situations {T,2} or {T,23}, when filling of first base will permit a force
play or (for T < 2) a double play on a subsequent ground ball.

The advisability of the strategy (aside from the factor associated
with the individual skills of the batters) can be examined by using the
data from Table V to compare probable scores following the situation {T,B}
before the walk with {T,B'} afterwards. This is done in Table VI.

There are several cases for which

$$\Delta = P(>r|T,B') - P(>r|T,B) < 0$$

indicating that the probability of scoring more than r runs has been reduced by issuing the walk. However, in order to determine whether the difference Δ is statistically significant, it must be compared to σ_Δ, the standard error of Δ, whose calculation is shown in [Lindsey, 1963]. We will consider a difference Δ to be not significant unless $|\Delta| > \sigma_\Delta$.

In Table VI, none of the differences Δ that are negative are as large (in absolute value) as σ_Δ, so that there is no clear statistical evidence that an intentional walk given to a batter in a team consisting only of average batters ever reduces the probability of scoring. In a number of cases Δ is positive and exceeds σ_Δ. These are indicated by the word 'Yes' in the column headed 'Δ', and represent cases in which a walk appears to be significantly detrimental to the chances of the fielding team.

It seems surprising that the walk with a man on second and one out should increase the probability of scoring at least one run. However, the difference $\Delta = P(>0|1,12) - P(>0|1,2)$ of .039, amounting to 1.4 σ_Δ on the basis of the pooled results from the 1959 and 1960 seasons, is -.002 for the 1960 season only.

A blank in the Δ column indicates that the difference is not significant according to the stated criterion. It will be observed that there are more blanks than Yesses, so that in the majority of these circumstances it cannot be determined whether the issuing of a base on balls to a member of a team of average batters is a good or a bad strategy. The wisdom of the strategy will turn mainly on the above- or below-average skill of the batter and of those who follow him in order.

Table VI

The Circumstances under Which the Issuing of a Base on
Balls Increases the Probability of Scoring

Before		After	$P(>0|T,B)$	$P(>0|T,B')$	Δ	$P(>1|T,B)$	$P(>1|T,B')$	Δ	$P(>2|T,B)$	$P(>2|T,B')$	Δ	$E(T,B)$	$E(T,B')$	Δ
B	T	B'												
2	0	12	.619	.605		.275	.385	Yes	.146	.254	Yes	1.194	1.471	Yes
2	1	12	.390	.429	Yes	.166	.266	Yes	.062	.147	Yes	.671	.939	Yes
2	2	12	.212	.209		.054	.109	Yes	.016	.048	Yes	.297	.403	Yes
23	0	F	.82	.83		.57	.56		.31	.39	Yes	1.96	2.254	Yes
23	1	F	.73	.69		.49	.448		.21	.262	Yes	1.56	1.632	
23	2	F	.332	.355		.237	.241		.067	.131	Yes	.687	.861	Yes

As stated previously, the decision regarding the advisability of a particular strategy must be made with regard to the state of the game. Suppose that the Home Team H are batting in the ith inning, and are ℓ runs ahead, with the situation $\{T,B\}$. The fielding team V can convert this situation into $\{T,B'\}$ by giving up an intentional base on balls. This move will increase their probability of winning the game if

$$\Delta = \Omega(H,i,\ell,T,B') - \Omega(H,i,\ell,T,B) < 0.$$

As before, the criterion for the difference to be statistically significant will be taken to be $|\Delta| > \sigma_\Delta$.

Table VII shows the results of such calculations for the last half of the ninth inning. In the row labelled 'Δ', the word 'Yes' signifies that $\Delta < -\sigma_\Delta$, and occurs for the case with men on second and third, one out, and the Home Team one run behind. As a matter of fact, $\Delta = -0.041$, and $\sigma_\Delta = 0.038$, and a difference of this sign and magnitude would be expected to be observed on 14 per cent of occasions between two random variables drawn from the same population.

It would seem that with less than two out, a man on third is quite likely to score on plays such as a sacrifice fly or a slowly hit ground ball. With men on second and third, a clean hit will score two runs and win the game immediately. Hence, with one out it is particularly important for the fielding team to maximize the probability of a game-ending double play. But without third base occupied the batting team has a much smaller probability of scoring the tying and winning run on one play. An intentional walk which may put the winning run on base is not advisable.

Table VII

The Circumstances Under Which the Issuing of a Base on Balls by V in H9 Increase the Probability that V will Win the Game

Situation before	B	2	2	2	2	2	2	2	2	2
Number Out	T	0	1	2	0	1	2	0	1	2
V Ahead by	ℓ	2	2	2	1	1	1	0	0	0
BBH helps V?	Δ	No	No	No	No	No	No			
Situation before	B	23	23	23	23	23	23	23	23	23
Number Out	T	0	1	2	0	1	2	0	1	2
V Ahead by	ℓ	2	2	2	1	1	1	0	0	0
BBH helps V?	Δ			No		Yes				

There are a number of cases in Table VII, indicated by blanks, for which $|\Delta| < \sigma_\Delta$; and a number marked by 'No' for which $\Delta > \sigma_\Delta$, indicating that the base on balls is significantly detrimental.

Similar calculations for the first half of the ninth inning show a large number of cases for which the change in probability of winning is less than σ_Δ, but only one for which the fielding team H can produce a significant increase in their probability of winning by issuing a base on balls. This instance occurs when the score is tied and, again, there are runners on second and third with one out. In most cases where H are ahead, they are distinctly ill-advised to give an intentional walk.

Tabulation of all the possible circumstances in all innings would be quite lengthy, and will not be attempted here. In general, it may be concluded that, aside from the personal abilities of the batters about to come up in the actual situations, there are not many circumstances in which an intentional walk improves the chances that the fielding team will win the game. The circumstances are most favorable for the strategy when the fielding team is ahead by one run and, as before, the situation is $\{1,23\}$.

The Circumstances Under Which a Sacrifice is Advisable. By the use of a skillful sacrifice bunt it is usually possible to advance all runners one base while the batter is put out at first. The laying down of a successful sacrifice bunt is not an easy matter, especially when it has been anticipated by the fielding side, and several undesired results may follow, such as a double play, or the elimination of a runner instead of the batter. It is also possible that everyone will be safe. However, for this study we will assume that the sacrifice will produce the intended result, and we will examine the desirability of this result in various circumstances. It should also be borne in mind that the batting team can adopt a mixed strategy, making it difficult for the fielding team to anticipate their actions.

The two usual reasons given for sacrificing a runner from first to second base are that he can probably score directly from second if a succeeding batter hits a single, and that the opportunity for the fielding side to make a double play is reduced. Such a sacrifice, if successful, converts the situation from $\{T,1\}$ to $\{T+1,2\}$. Comparison of $P(r|T,1)$ with $P(r|T+1,2)$ shows that while the probability of scoring exactly one run is increased, the probabilities of scoring at least one, of scoring exactly two, and of scoring more than two are all decreased. The expected score is decreased substantially. These statements are true both for T=0 and T=1, and all are statistically significant except for the probability of scoring at least one run with a man on first and none out. See Table VIII.

For the last half of the ninth inning it is found that $\Omega(H,9,\ell,T+1,2) < \Omega(H,9,\ell,T,1)$, for T=0,1, and all values of $\ell \leq 0$. The difference is less than σ_Δ for T=ℓ=0, but is significant for the other situations. This indicates that, with average batters coming up, the Home Team is not well advised to sacrifice with a runner on first base in the last of the ninth, except possibly with none out and the score tied.

In the first half of the ninth, there is no case for which a sacrifice by the Visiting Team with a runner on first base (and average batters coming up) makes a significant increase to their probability of winning the game. The sacrifice is a bad strategy when they are behind, or the score is tied with one out. When they are ahead, or the score is tied with none out, the difference is not significant.

Table VIII

The Situations in Which a Sacrifice
Increases the Probability of Scoring

Situation before		Situation after		$P(>0)$ Δ	$P(>1)$ Δ	$P(>2)$ Δ	E Δ
B	T	B'	T+1				
1	0	2	1		No	No	No
1	1	2	2	No	No	No	No
2	0	3	1	Yes	No	No	No
2	1	3	2	No	No	No	No
3	0	H	1	Yes	No	No	No
3	1	H	2	Yes	No	No	Yes
12	0	23	1	Yes	Yes	No	
12	1	23	2	No		No	No
13	0	2H	1	Yes	No	No	No
13	1	2H	2	Yes		No	Yes
23	0	3H	1	Yes	Yes	No	
23	1	3H	2	Yes	No	No	No
F	0	23H	1	Yes	Yes	Yes	Yes
F	1	23H	2	Yes	No		

Table VIII shows all of the situations in other innings in which a sacrifice could conceivably help the batting team. The entry 'H' in the column 'B'' indicates that a runner has scored. The entry 'Yes' in the table signifies that the probability of scoring more than r, or the expected score E, would be increased by more than σ_Δ by the execution of a sacrifice.

Most of the 'Yes' items occur in the column giving the probability of scoring one or more runs. Also, most of them are found for situations with a runner on third base. The placing of a bunt to bring a runner home from third base is the famous 'squeeze play', and is difficult to execute. If the batter misses the ball, the runner is very likely to be trapped between third and home. Thus, the possibility that the attempted sacrifice will produce an unsought result, which is neglected in these calculations, is particularly important in the cases with a runner on third base.

Base Stealing. In the previous section it was assumed that the chosen strategy could be executed. However it was not certain, and required calculation, to determine whether the successful carrying out of the strategy increased the probability of winning the game. Consideration of the advisability of having a runner attempt to steal a base is different. In this case there is no doubt that the advancement of the runner increases the probability of winning. But it certainly cannot be assumed that the attempt to steal will be successful.

The combined records of both major leagues for the years 1954-1955 and 1958-9 show 2,959 stolen bases and 2,057 instances when runners were caught stealing, so that the proportion of successes was 0.588. In 1971 the pro-

portion was 0.628. However, the success of individual players varies over
a wide range. The estimate of a runner's "true success rate" suffers from
all the difficulties of sampling errors as already discussed for batters,
and it seems evident that the probability of stealing successfully must be
very dependent on the style of the pitcher and arm of the catcher. There
are also auxiliary considerations, such as the effect of a long leadoff or
a past record of attempted steals on drawing throws from pitcher to first
base, distracting the pitcher, or motivating pitchouts or only fastballs,
to the probable benefit of the batter.

For this case, the decision rules for stealing bases can be set in
terms of the success rate needed to make the play advisable, instead of in
terms of the expected result based on average performance. The figures
quoted for success and failure include attempts to steal second, third, and
home base, and double steals. However, the vast majority are undoubtedly
single attempts to steal second base.

When a runner attempts to steal a base, he may succeed, he may be caught
stealing, or another play may occur (for example, if the batter hits the ball).
In this analysis we will consider only the first two possibilities, and compare the probabilities of scoring and of winning the game in the two alternatives that

(a) the attempt to steal succeeds, and converts the situation
$\{T,B\}$ into $\{T,B'\}$, and

(b) the runner is caught stealing, converting the situation
$\{T,B\}$ into $\{T+1,B''\}$.

Let S be the probability that the attempted steal will succeed. Then
the probability of scoring r runs by the end of the half-inning will be con-
verted by the decision to attempt a steal from $P(r|T,B)$ to $SP(r|T,B') +
(1-S)P(r|T+1,B'')$. This will be an increase if

$$S > [P(r|T,B) - P(r|T+1,B'')]/[P(r|T,B') - P(r|T+1,B'')] = S_c$$

where we define S_c as the critical probability of success above which the
probability of scoring will be increased.

Table IX shows S_c for the cases where runners are on first or first and
third. The symbol ~ 1 in a column headed $P(>r)$ signifies that according to
the data in Table V, completion of a successful steal will not increase the
probability of scoring more than r runs. It seems illogical that this should
really be the case. However, in all of the cases in Table IX for which

$$\Delta = P(>r|T,B') - P(>r|T,B) < 0$$

we have $|\Delta| < 2\sigma_\Delta$, so that these results can be attributed to sampling
error. For these cases it is concluded that a very high probability of
success (i.e., nearly 1.0) is required for the attempt to be profitable.

In examining Table IX we should remember that the probability of success in an attempted steal of second by an average runner is about 0.60 but
that the best runners achieve a much higher probability (0.7 to 0.8 or
higher).

Table IX

Critical Probabilities of Success if an Attempted
Steal is to Increase the Probability of Scoring

Situation before		Contested Steal	Critical probability of success S_c			
B	T		P(>0)	P(>1)	P(>2)	E
1	0	2nd	.53	.79	.66	.60
	1		.62	.83	.78	.70
	2		.54	~1	~1	.74
13	0	2nd	~1	.73	.88	.96
	1		.79	.41	.56	.63
	2		.85	.49	~1	.77

The situation in which a steal is most frequently attempted is {T,1},
i.e., a man on first base only, especially if he is a better-than-average
runner. The first three lines of Table IX indicate that this strategy may
be barely profitable if the runner is an average one and one run is badly
needed (i.e., to tie the score or go one ahead, especially near the end of
the game), but is not a good one if two or more runs are required.

With men on first and third, it would usually be advisable to attempt
the steal of second if the runner on third could come in to score, but if
this cannot be assumed (as it cannot be for baseball of a high standard) the
attempted steal is attractive only in the cases with one or two out and an
important requirement for two or more runs.

In default of knowledge of the probability of success of stealing
third or home, it is difficult to evaluate the advisability of making the
attempt. However, the data do show that the advisability of an attempt to
steal home when one run is badly needed increases sharply with the number
of men out. The critical probability of success with two out is only about
1/4 to 1/3. This also demonstrates that the third-base coach should be
bold in sending a baserunner on to home plate when there are two out and he
represents the tying or winning run.

Calculations of S_c for the ninth inning were made for the case {T,1}
of a runner on first base only. For the last half of the ninth, the only
situations for which S_c < 0.60 are with the score tied and less than two
out, but S_c < 0.70 when the Home Team is one run behind with any number
out. In general, the attempt to steal second near the end of the game
appears more advisable with none out than with one out, and distinctly in-
advisable with two out except when the batting team is one run behind.

A MEASURE OF BATTING EFFECTIVENESS

Batting effectiveness is usually measured by the batting average (H/AB),
slugging percentage (TB/AB), and runs batted in (RBI).

A new approach to the assessment of batting effectiveness could be based on three assumptions:

(a)   that the ultimate purpose of the batter is to cause runs to be scored

(b)   that the measure of the batting effectiveness of an individual should not depend on the situations that faced him when he came to the plate (since they were not brought about by his own actions), and

(c)   that the probability of the batter making different kinds of hits is independent of the situation on the bases.

The value of a hit toward the scoring of runs can be estimated from the data in Table V, since a hit which converts situation $\{T,B\}$ into $\{T,B'\}$ increases the expected number of runs by $E(T,B') - E(T,B)$. The relative frequency with which a batter is faced with each of the 24 possible situations $\{T,B\}$ can be estimated from the values of $N(T,B)$ in the first 24 rows of Table V. For the calculation it is assumed that runners always score from second or third base on any hit, go from first to third on 50 per cent of doubles, and score from first on the other 50 per cent of doubles. The result of these calculations is shown in Table X.

Thus, a home run increases the expected score, on the average, by 1.42 runs, about 3.1 times as much as a single. A base on balls or a hit batsman is worth 0.33 runs, or 0.72 times as much as a single. The weighting of $0:1:2:3:4$ used to compute the slugging percentage is not very different, but undervalues the walk and single. Using the value of hits from Table X, a batter's record can be converted into a "run producing average", normalized to the number of times at bat. Walks and times hit by pitcher were not included, on the basis that they are more attributable to the pitcher than to the batter.

Based on this criterion, the most effective batter, by a wide margin, was Babe Ruth, with an average of .270 runs produced per time at bat during his 22 years in the major leagues. Ruth also leads in lifetime runs batted in (2217) and slugging percentage (.690). Ty Cobb, who leads in lifetime batting average (.367) by a considerable margin, produced comparatively few extra base hits, and has a run producing average of .217, which is below Ted Williams (.253), Lou Gehrig (.251), Jimmy Foxx (.242), Hank Aaron (.227), and Dick Allen (.220).

Table X

The Value of Hits for the Scoring of Runs

Type of Hit	Walk	Single	Double	Triple	Home Run
Value	0.33	0.46	0.82	1.07	1.42
Ratio	0.72	1.00	1.80	2.34	3.12

AVERAGE AND UNUSUAL PLAYERS

Most of the calculations in this paper have been based on the performance of "average players". The theoretical reasoning deals with probability distributions obtained from observation of many hundreds of games. The
decision rules regarding such choices as issuing an intentional walk or
ordering a sacrifice bunt are calculated for an "average situation", with
an "average batter" coming up, followed in the order by a succession of
more "average batters". But in most real situations the manager may not
consider that the players concerned are average. Or he may order a re-
placement batter, runner, or pitcher, whom he believes to be above average
for the purposes desired. In principle, one could try to determine $P(r|T,B)$
for various batting averages, slugging percentages, or other measures of
the powers of the man due to bat. Even more suitable would be $P(r|T,B)$ for
combinations of the next two or even three men scheduled to bat.

The decision rules for base stealing do take account of individual
performance of the base runners. However, it is very difficult to make an
accurate estimate of the probability of a successful steal, since it certainly depends on the pitcher and catcher as well as the runner.

Perhaps the ideal manager should use these calculations of decision
rules as his first approximation, strictly valid only for average players.
His contribution would then be to judge the extent to which the key players
differ from the average, and to make an appropriate allowance for this. His
judgment regarding deviation from the average may be aided by detailed records kept for individual players and various circumstances, but there will
usually be some limits placed on the significance of this type of data because of the effects of sampling errors.

THE VALUE OF OUTSTANDING INDIVIDUALS TO THE TEAM

A feature of management unrelated to field strategy is the evaluation
of players for the purposes of trading and of salary agreement. Some of
the considerations extend beyond the realm of performance statistics, because they involve personality, amenability to discipline, box-office
appeal, or potential for future employment as coach, scout, or manager.
Moreover, next year's salary is likely to depend less on past achievements
than on the performance expected next year. Nevertheless, it is suggested
that a batter's "Run Producing Average" over the recent past would make one
good measure of effectiveness for assessing the probable value of a player
to the team in the near-term future, unless there are definite indications
that he is likely to improve or decline.

Another decision that must be made by the management is the apportion-
ment of their financial resources amongst the different positions on the
team. The largest division is between batters and pitchers. This could be
considered as offence and defence, but there is an interaction, since batters also field (and pitchers also bat, except in the American League since
1973).

It would be interesting to be able to determine whether a batter is
worth more than a pitcher. However, in its basic form the question is

meaningless. A team must have quite a few of each. A related question does offer the possibility of an objective answer: is a better-than-average batter of more value to the team than a better-than-average pitcher?

First, it is necessary to be more specific regarding the term "better-than-average". Let us select two measures of effectiveness, one for batters and one for pitchers, and observe the distribution in values recorded by individual players over the past season. Let us measure the variability among players by the standard deviation σ of these distributions, and define for the purposes of this calculation an "outstanding" batter (or pitcher) as one who achieves a score of $+\sigma$ above the mean performance.

Probably the best measure of effectiveness for pitchers is the Earned Run Average (ERA). Sixty-eight pitchers worked 100 or more innings in the National League season of 1973, representing 72% of all the innings pitched. The distribution of their individual ERAs had a mean value of 3.45 per nine-inning game and a standard deviation of $\sigma_p = 0.73$. One could therefore say that our "outstanding" pitcher "saved" 0.73 earned runs per nine-inning game for his team, as compared to the performance of an "average" pitcher.

Let us use our "Run Producing Average" (RPA) to translate the performance of the 1973 National League batters into runs contributed to their team. Ninety-seven men had 300 or more Times at Bat, representing 71% of all the Times at Bat in the season. For these 97 the RPA was calculated, and the distribution had a mean value of 0.169 and a standard deviation of $\sigma_B = 0.028$. One could say that an outstanding batter produced .028 extra runs for his team for each Time at Bat, as compared to the contribution of an "average" batter. Since the contributions of fielding errors in allowing the batter to reach base are not included, these extra runs can be considered as "extra earned runs", directly comparable to the earned runs "saved" by the pitchers.

To make a direct comparison of the contribution to the team's success of the outstanding batter and pitcher, these contributions of runs need to be normalized. The mean number of Times at Bat in a nine-inning game is about 34, so that those batters playing the full nine innings will average 3.8 Times at Bat. Hence a batter whose RPA exceeds the average by σ_B will contribute, on the average, $3.8\sigma_B = .106$ "extra" runs for each game in which he plays.

This could be interpreted to mean that an outstanding pitcher is more important than an outstanding batter in the ratio of .73/.106 = 6.9. However, the ratio applies only for the games in which both appear. This large ratio justifies the close attention paid to the expected pitchers by writers forecasting the results of games. It also supports the belief that "pitching is three-quarters of a short series". If we suppose that a good team possesses two outstanding pitchers and two outstanding batters, it may be possible to use the pitchers in 2 out of 3 games, contributing 2 x .73 = 1.46 runs "saved" in the series. The two outstanding batters will only contribute 3 x 2 x .106 = .63 "extra" runs. So the star pitchers make 1.46/2.09 = 70% of the contribution by outstanding players.

However, in considering the value of an outstanding player to the team (and consequently the money that should be offered to acquire or keep him), account must be taken of the fact that many batters play in nearly all of the games in a season, but no pitcher in more than one-quarter of the innings. The 97 batters with 300 or more Times at Bat averaged 481 Times at Bat for the season, and $481\sigma_B$ represents 13.5 "extra" runs for the season. The 68 pitchers with 100 or more Innings Pitched averaged 193 Innings Pitched for the season, equivalent to $193/9 = 21.4$ nine-inning games, and $21.4\sigma_p$ represents 15.7 runs "saved" for the season. Thus, once the "utilization factor" is included, the contributions for the season of a batter and a pitcher, each exceeding the average by one standard deviation, and each appearing in the average number of innings, was not very different.

This brings up the importance of utilization. It is inherent in certain measures of effectiveness of batting, such as home runs or runs batted in, but is factored out of measures such as batting average, slugging percentage, RPA, or ERA. A case could be made for a measure of effectiveness that multiplied the deviation from the mean by utilization: for batters it would be (RPA-$\overline{RPA}$) (AB), and for pitchers (ERA-$\overline{ERA}$) (IP)/9, where the bars represent the mean for all players, or for all players participating above a certain minimum level (such as 300 Times at Bat or 100 Innings Pitched). The units would be "extra" runs and runs "saved" per season, with zero representing "average" performance. About half the players would have a negative index.

Table XI summarizes these results for the National League for 1973 and also shows the performance of several unusual individuals. Willie Stargell, who had the best Slugging Percentage (.646), most runs batted in (119), and most home runs (44), contributed 42.8 "extra" runs, the most of any player by a considerable margin. Stargell's RPA of .251 exceeds the mean RPA by $2.9\sigma_B$ and he appeared in over 90% of Pittsburgh's games. Pete Rose, who had the best batting average (.338), most hits (230), and most Times at Bat (680) had a small number of "extra" runs. This was because 79% of Rose's hits were singles (as compared to 42% for Stargell). For the pitchers, Tom Seaver combined a very low ERA with a very high number of Innings Pitched, to "save" 44.1 runs, more than Stargell produced as "extras". Seaver was far ahead of all other pitchers. Tommy John, with the highest Won-Lost Percentage (16-7), had an ERA close to the average of 3.45 and Ron Bryant, with the most games won (24-12) had an ERA above the average, and consequently records a negative score for runs "saved".

A conclusion of this analysis could be that an outstanding pitcher has far more influence than an outstanding batter for those games in which he pitches, but over the entire season there is less imbalance between the contribution of outstanding pitchers and outstanding batters. However, a really outstanding pitcher able to work nearly 300 innings in a season is likely to be worth more to the team than almost any outstanding batter.

Table XI

Extra Runs Contributed by Outstanding Batters and
Runs Saved by Outstanding Pitchers, National League, 1973

Batters (≥ 300 AB)

	AB	RPA	"Extra" Runs
+1σ_B Batter			
per Time at Bat	1	.197	0.028
per 9-inning Game	3.8		0.106
per Season	481		13.5
Average Batter	481	.169	0.0
1. Stargell (.299)	522	.251	42.8
2. Aaron (.301)	392	.248	31.0
3. Evans (.281)	595	.218	29.1
Rose (.338)	680	.188	12.9

Pitchers (≥ 100 IP)

	IP	ERA	Runs "Saved"
+1σ_P Pitcher			
per 9-inning Game	9	2.72	0.73
per Season	193		15.7
Average Pitcher	193	3.45	0.0
1. Seaver (19-10)	290	2.08	44.1
2. Sutton (18-10)	256	2.43	29.0
John (16-7)	218	3.10	8.5
Bryant (24-12)	270	3.53	-2.4

ACKNOWLEDGMENTS

The author would like to acknowledge the cooperation of the editors
of Operations Research and of the Journal of the American Statistical Society,
for permission to reproduce diagrams, tables, and passages of text. In
addition, thanks and recognition are due to the late Lieutenant Colonel
Charles Lindsey, who collected the majority of the data, and to Robert E.
Machol, for helpful suggestions regarding analysis and presentation.

BATTING PERFORMANCE VS. BALL-STRIKE COUNT

by

Pete Palmer

In order to analyze the effect of ball-strike count on batting performance in baseball, pitch-by-pitch data was taken for twelve World Series games in 1974 and 1975. Batting results were compared using weights for each result based on the equivalent number of runs produced above average. These values, derived from a previous study, are: 0.50 for a base hit, 0.30 for each extra base, 0.33 for each base on balls or hit batsman and -0.25 for each out. (A home run is thus worth 1.4 runs, for example.) The table below shows the results for all batters who passed through the count indicated. For example, if the batter took a ball, then swung and missed, and then singled, the single would be included under 0-0, 1-0, and 1-1 counts. Sacrifice bunts and flies are excluded from the data, as are intentional bases on balls. In these games, the average number of runs per faced pitcher (R/FP) was 0.10 (the numbers shown in the table are differences from this average). The pooled data for each B-S category (Average) shows that a batter one pitch ahead on the count does 40% better than average, while a batter one pitch behind does 40% worse than average. There is a definite linear relationship between batting performance and balls minus strikes across the full range of count possibilities. Since each batter success is worth about 0.55 runs and each failure is worth -0.25 runs, the net gain for an additional success is 0.80 runs. Thus one standard deviation in R/FP for any sample is $\sqrt{P \times (1-P)/FP} \times 0.80$ where P is (BH+BB)/FP. The differences in the table are in the 0.95 significance range.

The study shows the importance of getting the ball over the plate for the pitcher and the value of not swinging at bad pitches for the batter. The degree of effect of count on performance seems rather greater than one might expect. There are also strategic implications: "pitching out" when a steal is anticipated and "wasting" a pitch on a count of 0-2 should be used more sparingly. I thank Robert E. Machol for help with this article.

B-S	COUNT	FP	AB	BH	TB	BB	R	R/FP	Difference in AVERAGE
3	3-0	50	13	6	6	37	+14	+.28	+.28
2	3-1	79	41	12	20	38	+14	+.18 }	+.13
	2-0	138	88	20	32	50	+14	+.10 }	
1	3-2	104	64	10	10	40	+ 5	+.04)	+.04
	2-1	172	143	33	55	29	+ 5	+.03 }	
	1-0	363	295	74	118	68	+17	+.05)	
0	2-2	157	132	30	44	25	+ 2	+.01	.00
	1-1	322	294	72	100	28	- 2	-.01	
	HFP	142	140	36	54	2	- 2	-.01	
	0-0	867	783	185	284	84	0	.00)	
-1	1-2	211	197	43	54	14	- 9	-.04 }	-.04
	0-1	362	348	75	112	14	-14	-.04 }	
-2	0-2	133	131	22	27	2	-14	-.10	-.10

B-S = balls minus strikes   BH = base hits   TB = total bases
COUNT = count in balls and strikes   FP = total faced pitcher (AB+BB)
AB = times at bat   BB = bases on balls (includes hit batsmen)
BH = base hits   AVERAGE for this (B-S), weighted by number in FP column
R = runs above average = $0.50 \times BH + 0.3 \times (TB-BH) + 0.33 \times BB - 0.25 \times (AB-BH)$
HFP = batters who hit the first pitch (plus 2 who were hit)

THE COMPUTER-ASSISTED HEURISTIC APPROACH
USED TO SCHEDULE THE MAJOR LEAGUE BASEBALL CLUBS

by

William O. Cain, Jr.

The purposes of this article are to present a problem and to describe the approach used on the problem in practice. Since the 1970 season, the author has made the major league baseball schedules under the direction of Mr. Fred G. Fleig, the National League Secretary, and Mr. Robert F. Holbrook, his American League counterpart.

The Structure of the Schedule. The 162-game schedule which has been played since the expansion to twelve clubs in each league in 1969 provides for each club playing eighteen games with each of the five other clubs in its own division (5 x 18 = 90) and twelve games with each of the six clubs in the other division (6 x 12 = 72).

The basic unit of the schedule is the three-game series. With each opponent, a club will have equal numbers of home and road games. So, in terms of series, there are fifteen home and fifteen road intradivisional series and twelve home and twelve road interdivisional series. At the time this article was written Baltimore, Boston, Cleveland, Detroit, Milwaukee, and New York were in the American League Eastern Division; while California (L.A.), Chicago, Kansas City, Minnesota, Oakland, and Texas were in the West. The National League East consisted of Chicago, Montreal, New York, Philadelphia, Pittsburgh, and St. Louis; and the West had Atlanta, Cincinnati, Houston, Los Angeles, San Diego, and San Francisco.

Normally, a three-game series is played either over a weekend on Friday, Saturday, and Sunday, or during the week on Tuesday, Wednesday, and Thursday. Since a total of 54 three-game series is required for the 162-game schedule, it would take 27 weeks to play them all at a rate of two series a week. In recent years, the season has been about 25 weeks long. A further complication has been the three-day break for the All-Star Game, which further reduces the playing dates available.

The accommodation made to deal with this situation is to introduce several "squeeze weeks" into the schedule. During the squeeze week, three series are played: two two-game series on Monday and Tuesday and on Wednesday and Thursday, and the usual three-game weekend series. The playing of a two-game series with an opponent forces a club to play a four-game series with that opponent at some other time. The All-Star Game is usually on a Tuesday, with off-days immediately before and after. In the five days (Thursday-Monday) after the break, two series are usually played, one of which will be two games. At this time, a club will often choose to play a Sunday double-header to get a third game in.

Desiderata. The easiest characteristic of a schedule to measure is the total travel cost required, which is almost exactly a linear function of mileage. As a crude rule-of-thumb, it is usually assumed that travel expenses for a club will be $4 per mile. The total mileage for both leagues averages about 750,000 miles a year. Hence total annual travel expense is about $3,000,000.

A much more difficult aspect of the schedule to evaluate is the expected attendance it will generate. Total attendance for 1974 was about 30,000,000 fans. Using $3.50 as an approximation of the average net revenue from one fan from his ticket, concessions, and parking, one can compute that $100,000,000 would approximate total receipts. In 1970, total attendance was about 28,750,000. The difference in receipts between the two years would be about $4,000,000. These figures demonstrate that the effect of a schedule on attendance is much more important than its effect on mileage. In the absence of a predictive model for attendance, the leagues proscribe certain features felt to be detrimental to attendance and insist on certain other features which may enhance attendance.

One proscribed feature is the "back-to-back", or the playing of consecutive series between the same two clubs. "Semirepeaters", the playing of two series between the same two clubs with only one intervening series, are discouraged, but not proscribed.

Another aspect of the schedule having an effect on attendance is the length of home stands and road trips. These should be neither too long nor too short. Home stands that are too short prevent any build-up of fan interest. Home stands that are too lengthy deplete both interest and finances. Road trips that are too long may undercut interest in going out to the ball park. In general, the leagues regard four series (two weeks) as the maximum desirable length for either a home stand or road trip, and two series (one week) as the minimum desirable length for a homestand. One-series road trips are permissible, since they are not believed to hurt attendance, but more than one or two per year for any club is considered bad scheduling because of the inconvenience caused the players.

The leagues insist upon good summer weekend series between natural rivals such as the Dodgers and the Giants (especially in San Francisco) and the Cardinals and the Cubs (especially in Chicago).

Both leagues have clubs in the New York, Chicago, San Francisco, and Los Angeles areas. In order to avoid splitting attendance between two clubs, the leagues insist that both clubs in a "conflict city" not be home at the same time. The effect of this is that at any one time exactly one of the clubs is home. Thus, anyone in one of these cities will always have an opportunity to see a major-league baseball game at any time during the season.

Besides mileage and attendance, there is a third aspect in the evaluation of the schedule: fairness. This is important because each league is composed of twelve independent franchises. Although the visiting club gets a share of the gate receipts, most of the receipts go to the home club. Certain features must be present in the schedule in order that it be reasonably fair.

Since weekend series are important, each club should have the same number of home weekends as road weekends. Further, since some visiting clubs will draw better than others, every club must visit every other club on a weekend. This implies that every club must have at least one home weekend series with each of the eleven other clubs. This requirement is particularly hard to satisfy because each club will have only twelve or thirteen home weekends in a 25-week season.

Attendance during the summer vacation months being better than attendance while schools are in session, the leagues insist that all clubs have an approximately even distribution of home games by month. Since there are twenty-eight playing days in a typical month, this means that each club should normally have at least four home series (twelve games) in every month.

While it is true that all clubs in a division play the same number of games with each opponent, it would be considered bad scheduling for one club to have a disproportionate number of games in a short period of time with either strong or weak opponents. Putting this another way, the schedule should have no perceived effect on the pennant races.

Besides these general considerations, there are also considerations specific to individual clubs. Local holidays, such as Jean-Baptiste and Dominion Days in Montreal, and Patriots' Day in Boston, offer a chance to increase attendance. Over the years, every club (except Montreal) should be home on the Fourth of July about half the time.

Every year, several clubs also make specific requests not to be at home on certain days. Billy Graham may be using the stadium, or some competing event may be in town. Some clubs share stadiums with professional football teams in the autumn.

The Home-and-Away Pattern. In the approach discussed in this paper, the home-and-away pattern plays an elemental role. As the name suggests, it is simply an indication of which clubs are home and which away during each series and does not show when a club visits any particular club. As we shall see, however, there may be only one opportunity for one club to visit another. If such is the case, it will be true that the pattern forces certain visiting assignments to be made.

To demonstrate how the home-and-away pattern affects the schedule, consider first a small example. It involves just four teams: A, B, C, and D. The "season" considered consists of six series. This allows each club to play one home and one road series with each opponent. A possible home-and-away pattern for such a schedule is shown in Fig. 1a, with circles representing home series and dashes representing away series.

There are two features in this pattern that are characteristic of any pattern in which all clubs play one home and one road series with every opponent: the number of circles in each column is one less than the number of clubs; and, in any one row the number of circles is half the number of clubs.

Figure 1.   A Home-and-away Pattern and Its Associated Schedules

```
      A B C D            A B C D            A B C D
  1   0 0 - -        1   D C - -        1   C D - -
  2   0 - 0 -        2   B - D -        2   D - B -
  3   0 - - 0        3   C - - B        3   B - - C
  4   - 0 - 0        4   - A - C        4   - C - A
  5   - 0 0 -        5   - D A -        5   - A D -
  6   - - 0 0        6   - - B A        6   - - A B
        Fig. 1a            Fig. 1b            Fig. 1c
```

While a pattern does not determine a unique schedule, it permits only a small number of feasible solutions when two simple rules are applied. The first rule is that if there is only one opportunity for one club to visit another (i.e. when one club is on the road, not already assigned to visit someone else, and another club is at home, but not being visited by anyone else), it must make its visit at that time. The second rule is that if one of the two away clubs has been assigned to visit one of the two home clubs during a particular series, then the unassigned away club must visit the remaining home club during that same series.

In the pattern shown in Fig. 1a, only during series 2 and 3 does B have an opportunity to visit A. If we choose to have B play at A during Series 2, application of the two rules above shows the only feasible schedule to be the one in Fig. 1b. If B plays at A during series 3, only Fig. 1c is a feasible schedule.

Thus, we see that making just one choice determined the entire schedule in this case. The two schedules in Fig. 1b and 1c are the only two that can be made from the pattern shown in Fig. 1a.

Other types of patterns exist which actually force certain assignments to be made because the pattern affords only one visiting opportunity for a particular club in a certain city. After these forced choices have been made, the procedure described above can be used to determine all schedules having a particular pattern.

This section has demonstrated in miniature how to construct a schedule given a pattern. A similar approach is used to construct schedules from a 6 x 10 pattern. With six clubs in each division, each club has five opponents. Playing one home and one road series with each opponent gives a total of ten series. This might also be called the "September problem" because each division plays such a schedule the last five weeks of the season. Because they do, the divisional championships are decided by intradivisional play.

An important element in the making of the whole schedule is the construction of a 6 x 10 pattern and the generation of the best schedule having that pattern. The technique employed is an exhaustive search using a branching structure based on the fact that only a few choices need be made to determine a schedule. A computer program which does this is used frequently in the actual construction of the schedule.

The Approach Used in Practice. To decompose the problem into manageable pieces, a structure is imposed on the schedule. The first aspect of the structure is the division of the season into three phases. During the first two of these phases, each team plays every other team in the league (including those in the opposite division) two series, one at home and one on the road. Thus, the first phase is twenty-two series long, and ends about the middle of June. The second phase is the same length, ending in late August. In September, each club plays two series, home and road, with each of the five other clubs in its own division. The division of the season into such phases tends to distribute games with each particular opponent over the whole season. It also ensures that there will be two points during the season when all clubs will have played exactly the same number of series with all opponents.

Each of the first two phases is further subdivided into segments of

intradivisional and interdivisional play. The simplest subdivision used
has been putting ten series of intradivisional play at the beginning of
each phase and twelve series of interdivisional play at the end. This
creates a "five-layer cake" schedule consisting of alternating segments:
intra-, inter-, intra-, inter-, intra-. In this approach, intradivisional
opponents are played in April, June-July, and September. There are two
six-week interdivisional segments, one of which runs from the middle of
July through the end of August, the best-drawing part of the season.

Since intradivisional opponents generally draw better than interdivisional opponents, in recent years the schedule has been structured to
split up interdivisional play in the summer. For example, in the 1975
schedule, a "seven-layer cake" format is used. The first phase has the
usual two segments of intra- and interdivisional play. The second phase
consists of four alternating segments: intra-, inter-, intra-, inter-.

Once the arrangement of intra- and interdivisional segments is decided upon, the next step is the placement of the "squeeze weeks." These
consist of three series instead of the usual two series. How many squeeze
weeks will be necessary depends on the length of the season.

The most important effect of the placement of the squeeze weeks is on
the distribution of weekend series. Since the interdivisional schedules
in each phase are usually similar, placing an odd number of squeeze weeks
during the intradivisional play that intervenes staggers the interdivisional play, so that the interdivisional games that fall on a weekend in
the first phase will fall during the week in the second phase, and vice
versa. Thus, every club will have one home weekend series with each interdivisional opponent. It is important in placing the squeeze week in
the midsummer intradivisional play to avoid eliminating any intradivisional weekends during this time.

By far the most difficult part of the schedule is that involving the
California teams. Both leagues have clubs in the San Francisco Bay Area
and the Los Angeles area, and every attempt is made to avoid having both
clubs home simultaneously in either city. Further, the distances from California to all other cities are large relative to the distances among all
other clubs. For example, it is further from Los Angeles to the nearest
non-California city (Kansas City) than from Philadelphia to the farthest
non-California city (Houston). Finally, the Los Angeles Dodgers have been
consistently good gate attractions in other cities, and are especially important to San Francisco and San Diego. In the American League, the Oakland A's have drawn well on the road recently.

In the interest of fairness, the leagues alternate from year to year
in opening the season in San Francisco and Los Angeles. In 1974, the Dodgers and Giants from the National League opened at home; in 1975, it is the
A's and the Angels from the American.

The distance from California to the other cities makes mileage an important consideration in scheduling the California clubs outside California and the other clubs into California. As far as the California clubs
go, about the best that can be done is to limit them to nine trips outside
the state during the season. (Obviously, there could be fewer if longer
road trips were used, but this is considered to be even more undesirable
than the extra mileage.) The visits of Eastern Division clubs to California are made in groups of three for the National League clubs and in groups

of two for the American League clubs. Of course, while Eastern Division
clubs in one league are visiting California, the California clubs in the
other league are visiting Eastern Division clubs in their league. A typical arrangement of interdivisional play might have one pair of American
League Eastern Division clubs visiting Oakland and California; followed by
three National League Eastern Division clubs visiting San Francisco, Los
Angeles, and San Diego; then two pair of American League East clubs in Oakland and California (giving the Giants, Dodgers, and Padres a four-series
road trip); and finally the last three National League East clubs in California. Replicating this construction in the second phase enables all Eastern Division clubs to make their required visits to California in two trips.
The California clubs make four trips into the Eastern Division.

The next phase is the intradivisional scheduling for the Western Divisions of each league. This is made more difficult by the necessity for the
California teams to play each other. It is usually not possible for every
Western Division club not based in California to play every team in California on every visit to the state. When a team cannot do so, it must travel
to California twice during a single intradivisional segment. By rotating
this double trip among the Eastern Division clubs over the course of the season, the number of trips to California for any one club can be held to four
in the American League. Some National League clubs make five such trips.

As the Western Division schedules are developed, it is necessary to
keep in mind several considerations. Each California club must have a home
weekend series with every other club. The dates on which the Dodgers visit
the Giants and Padres should be good ones. No California team should have
more than nine trips out of the state. Holidays should be alternated between the leagues in the conflict cities. Finally, the distribution of series among the months must be equitable.

It is in making the intradivisional segments that the computer program
mentioned earlier is most useful. Given a home-and-away pattern it will
print out either all possible schedules or just the one with the least mileage. It also counts the number of semirepeaters. Recently it has been modified so that it will not generate schedules having back-to-backs. Since
some of the intradivisional segments lie between interdivisional segments,
the program accepts input giving the starting and ending locations for each
club. Since some intradivisional segments are interrupted by interdivisional play or the All-Star Game, it is also possible to specify when the interruption occurs and where the clubs go during the interruption. Sometimes it
is desirable to specify a certain pairing of opponents at a certain time;
the program permits this to be done as well. It is during this portion of
the schedule-making process that most of the man-machine interaction takes
place.

The completion of the Western Division schedules provides the first opportunity for the League Secretaries to criticize the way the schedule is
taking shape.   My response to their criticisms is to try to satisfy them
without going any further back in the schedule-making process than necessary.
This means trying to accommodate their criticisms first by changes in the intradivisional schedule, then in the interdivisional schedule, and--only if
these fail--in the arrangement of the squeeze weeks, the most basic element.

The major thing the League Secretaries are likely to criticize is the
allocation of June, July, and August home dates to each club. They will

also object if the same club opens with another one two years in a row. Because the visiting teams share in the gate receipts, it is desirable for all clubs to get a "road opener:" that is, to be the visiting club for someone else's Opening Day. These are less important considerations than avoiding excessive travel or unbalanced June, July, and August dates, however.

The next phase is the completion of the interdivisional schedules. Here, the most important consideration is that they fit with the intradivisional schedules in such a way as to make a reasonable home schedule for each club over the entire season.

The final step is the preparation of the Eastern intradivisional schedule. Since the Chicago White Sox of the American League are in the Western Division, the pattern of home-and-away dates for the Chicago Cubs in the National League Eastern Division is already determined. Indeed, with the interdivisional segments also done, there is little freedom in the determination of the pattern. The Yankees and Mets alternate opening the season in New York, as do the White Sox and Cubs in Chicago.

Now is when the most important interaction with the League Secretaries takes place. They examine the draft, criticizing each club's schedule. Again, my response is to try to correct the undesirable features as simply as possible. Sometimes a simple switch of two series will suffice; more often, it is necessary to return to one of the earlier portions of the construction process. As the refinement proceeds further and further, improvement becomes progressively more difficult. As this happens, it becomes increasingly necessary to make hard choices to give up one feature in order to attain another. About 500 man-hours and five hours of CPU computer time on a Univac Spectra 70/7 are spent in making the schedule.

Comparison of the 1969 and 1975 Schedules. To get an idea of the differences in results obtained using the computer-assisted approach and that used previously, consider the 1969 and 1975 schedules. The 1969 schedule is the most recent schedule not made by the author. (It was, incidentally, the first schedule made with twelve teams in each league.) The calendars for the two years coincide, so that the opening days are identical: April 7 in Cincinnati and April 8 in the other cities. The 1969 season ended on Wednesday, October 1, while the 1975 season ended on Sunday, September 28, so it required an extra squeeze week.

The National League schedules are presented for comparison rather than the American League schedules because there have been no franchise shifts in the National League. Therefore, it is possible to compare meaningfully the mileage required by each schedule. Further, because the National League Western Division includes two cities in the Eastern time zone, Cincinnati and Atlanta, it is more difficult to schedule than the American League. The 1969 and 1975 National League Western Division schedules by series may be found in Tables I and II.

Except for the number of semirepeaters (56 in 1975 vs. 46 in 1969), the 1975 schedule is superior in almost all respects. There are no five-series home stands in 1975, while there are four in 1969, including one in San Diego June 27 - July 13 that lasts for two-and-a-half weeks and includes three consecutive weekends. It is considered bad scheduling for a club to be either home or away for three consecutive weekends: there are ten instances of this undesirable feature in the 1969 schedule and none in the 1975 schedule. There were five two-and-a-half-week road trips in 1969,

compared to one in 1975; and two three-week road trips in 1969 compared to
none in 1975. Finally, the series were much more evenly distributed by
month in the 1975 schedule. The most glaring manifestation of this feature
is that four clubs have only three home series in July 1969. Such imbalance might be alleviated if these clubs had extra August series, but such
is not the case. In the 1969 schedule Atlanta, Cincinnati and Houston have
only seven home series in the months of July and August combined.

In both schedules, every club in the League has one home Sunday with
every other club. Further both schedules have four Sundays with Chicago
vs. St. Louis, two in each city, just as they do with Los Angeles vs. San
Francisco.

Except for the number of semirepeaters, all of the respects on which
the two schedules have been compared are really characteristics of the home-
and-away pattern itself. Focusing on the pattern makes it possible to
avoid many of the irregularities found in the 1969 schedule. At the same
time, having the computer available to search over all schedules for given
patterns makes it possible to keep the mileage reasonable.

The total mileage is virtually identical for both schedules, with 455
or 0.12% more mileage in 1975. The difference is in how it is distributed
among the clubs. Every Eastern Division club had slightly less travel in
1975 than in 1969. The mileage for the Western Division clubs is shown below:

	Atlanta	Cincinnati	Houston	Los Angeles	San Diego	San Francisco	Division
1969	26,276	27,980	38,935	41,267	41,784	43,220	219,462
1975	33,723	31,911	35,346	44,713	38,311	46,319	230,323
Dif-ference	+7,447	+3,931	-3,589	+3,446	-3,473	+3,099	+10,861

Atlanta, Cincinnati, Los Angeles, and San Francisco all had more travel in
1975 than in 1969, but each had compensating advantages. Atlanta and Cincinnati had nine home series, exactly the average, in the critical months
of July and August, instead of seven. Los Angeles did not have a three-
week span during these months without a home game. Finally, San Francisco
did not have two two-and-a-half week road trips. All other clubs had less
mileage in 1975 than in 1969 in addition to relief from the irregularities
discussed above.

The 1977 schedule is entirely different from previous schedules. The
American League is expanding to fourteen clubs. With seven clubs in each
division, it will never be true that any segment consists exclusively of
intradivisional play. Thus, it will not be possible to have the kind of
final month of intradivisional play that went on from 1969-1976. As of
this writing, it appears that American League clubs will play two home and
two road series with each of the thirteen other clubs, for a total of 52
series. However, three series with each intradivisional opponent will be
four games, for a total of fifteen games per opponent; two of the interdivisional series will be only two games, for a total of ten games per opponent. As a result of these departures from the three-game series, there
will be more intradivisional games (6 × 15 = 90) than interdivisional

Table I

1969 National League Western Division Schedule by Series

Sun.	Atl	Cin	Hou	L A	S D	S F
	S F	L A	@S D	@Cin	Hou	@Atl
4/13	Cin	@Atl	L A	@Hou	S F	@S D
	@Hou	S F	Atl	S D	@L A	@Cin
4/20	@Cin	Atl	@L A	Hou	@S F	S D
	@S D	@Hou	Cin	S F	Atl	@L A
	@S F	@L A	S D	Cin	@Hou	Atl
4/27	@L A	@S D	@S F	Atl	Cin	Hou
	Hou	@S F	@Atl	@S D	L A	Cin
	S D	Hou	@Cin	@S F	@Atl	L A
5/4	L A	S D	S F	@Atl	@Cin	@Hou
	Mtr	@N Y	@Pha	@Chi	@Pit	@StL
5/11	@Pha	@Mtr	@N Y	@Pit	@StL	@Chi
	@N Y	@Pha	@Mtr	@StL	@Chi	@Pit
5/18	@Mtr	N Y	Chi	Pit	StL	@Pha
	N Y	Pha	Mtr	Chi	Pit	StL
5/25	Pha	Mtr	N Y	StL	Chi	Pit
	StL	Pit	Pha	@Mtr	@N Y	Chi
6/1	@Chi	@StL	@Pit	@Pha	@Mtr	@N Y
	@StL	@Pit	@Chi	@N Y	@Pha	@Mtr
6/8	Pit	@Chi	StL	Mtr	N Y	Pha
	Chi	StL	Pit	Pha	Mtr	N Y
6/15	@Pit	Chi	@StL	N Y	Pha	Mtr
	Hou	@S F	@Atl	S D	@L A	Cin
6/22	S F	L A	S D	@Cin	@Hou	@Atl
	L A	S D	S F	@Atl	@Cin	@Hou
6/29	@Hou	S F	Atl	@S D	L A	@Cin
	Cin	@Atl	@L A	Hou	S F	@S D
7/6	@S F	@L A	@S D	Cin	Hou	Atl

Sun.	Atl	Cin	Hou	L A	S D	S F
	@L A	@S D	@S F	Atl	Cin	Hou
7/13	@S D	@Hou	Cin	S F	Atl	@L A
	@Cin	Atl	L A	@Hou	@S F	S D
7/20	S D	Hou	@Cin	@S F	@Atl	L A
*********	ALL	-	STAR	GAME	*********	
7/27	Mtr	@N Y	Pha	@Chi	@Pit	@StL
	@Pha	Mtr	@N Y	@Pit	@StL	@Chi
8/3	@N Y	@Pha	@Mtr	@StL	@Chi	@Pit
	@Mtr	N Y	Chi	Pit	StL	@Pha
8/10	N Y	Pha	Mtr	Chi	Pit	StL
	Pha	@Mtr	N Y	StL	Chi	Pit
8/17	StL	Pit	@Pha	@Mtr	@N Y	Chi
	@Chi	@StL	@Pit	@Pha	@Mtr	@N Y
8/24	@StL	@Pit	@Chi	@N Y	@Pha	@Mtr
	@Pit	@Chi	@StL	Mtr	N Y	Pha
8/31	Chi	StL	Pit	Pha	Mtr	N Y
	Pit	Chi	StL	N Y	Pha	Mtr
9/7	@Cin	Atl	S F	@S D	L A	@Hou
	L A	S F	S D	@Atl	@Hou	@Cin
	S F	S D	L A	@Hou	@Cin	@Atl
9/14	Hou	@S F	@Atl	S D	@L A	Cin
	@S F	@L A	@S D	Cin	Hou	Atl
	@L A	@S D	@S F	Atl	Cin	Hou
9/21	@S D	@Hou	Cin	@S F	Atl	L A
	@Hou	L A	Atl	@Cin	S F	@S D
9/28	S D	Hou	@Cin	S F	@Atl	@L A
	Cin	@Atl	@L A	Hou	@S F	S D

Away games denoted by @.

games (7 x 10 = 70), even though there will be more interdivisional series. The National League will probably expand in 1978.

Looking back, we see that eight-club leagues existed for eighty-six years (1876-1961); ten-club leagues for seven years (through 1968); and twelve-club leagues for eight years (through 1976). How long fourteen-club leagues will last is anybody's guess. If and when the leagues expand to sixteen teams each, it seems obvious that there will be in effect four eight-team leagues. Since these could be scheduled independently, the scheduling problem would be much simpler. The possibility of expansion makes the development of a more sophisticated computer scheduling model uneconomic at this time.

Acknowledgments. Fred Fleig introduced me to the baseball scheduling problem. I thank him and Bob Holbrook for their patience and encouragement, two essential requirements for supervisors of schedulemakers. I also thank Robert E. Machol for help with the presentation of this article.

Table II

1975 National League Western Division Schedule by Series

Sun.	Atl	Cin	Hou	L A	S D	S F	Sun.	Atl	Cin	Hou	L A	S D	S F
	@Hou	L A	Atl	@Cin	S F	@S D		N Y	Pha	Mtr	@Pit	@Chi	@StL
4/13	@S F	@S D	L A	@Hou	Cin	Atl	7/13	Mtr	N Y	Pha	@StL	@Pit	@Chi
	Hou	@L A	@Atl	Cin	@S F	S D	********* ALL - STAR GAME *********						
4/20	S D	Hou	@Cin	S F	@Atl	@L A		@N Y	@Mtr	@Pha	Pit	Chi	StL
	L A	S F	S D	@Atl	@Hou	@Cin	7/20	@Mtr	@Pha	@N Y	Chi	StL	Pit
	Cin	@Atl	S F	@S D	L A	@Hou		@Pha	@N Y	@Mtr	StL	Pit	Chi
4/27	@S D	@Hou	Cin	@S F	Atl	L A	7/27	S D	L A	S F	@Chi	@Atl	@Hou
	@L A	@S F	@S D	Atl	Hou	Cin		L A	S F	S D	@Atl	@Hou	@Cin
5/4	@Cin	Atl	@S F	S D	@L A	Hou	8/3	@S D	@L A	@S F	Cin	Atl	Hou
	S F	S D	@L A	Hou	@Cin	@Atl		@L A	@S F	@S D	Atl	Hou	Cin
5/11	Pha	@N Y	@Mtr	@Pit	@Chi	@StL	8/10	Chi	Mtr	Pit	@N Y	@StL	@Pha
	@Mtr	@Pha	@Chi	@StL	@Pit	@N Y		Pit	Chi	StL	@Pha	@N Y	@Mtr
5/18	@Pha	@Mtr	N Y	Pit	Chi	StL	8/17	StL	Pit	Chi	@Mtr	@Pha	@N Y
	Mtr	N Y	Pha	Chi	StL	Pit		Pha	@StL	N Y	@Chi	@Mtr	@Pit
5/25	N Y	Pha	Mtr	StL	Pit	Chi	8/24	@StL	@Pit	@Chi	Mtr	Pha	N Y
	@Chi	Mtr	@Pit	@N Y	@StL	@Pha		@Pit	@Chi	@StL	Pha	N Y	Mtr
6/1	@Pit	@StL	@Pha	@Chi	@N Y	@Mtr	8/31	@Chi	StL	@Pit	N Y	Mtr	Pha
	@StL	@Pit	@N Y	@Mtr	@Pha	@Chi		@Hou	S D	Atl	@S F	@Cin	L A
6/8	@N Y	Chi	StL	@Pha	@Mtr	@Pit		S D	L A	S F	@Cin	@Atl	@Hou
	StL	Pit	Chi	Mtr	Pha	N Y	9/7	L A	S F	S D	@Atl	@Hou	@Cin
	Chi	StL	Pit	N Y	Mtr	Pha		Hou	@S D	@Atl	S F	Cin	@L A
6/15	Pit	@Chi	@StL	Pha	N Y	Mtr		@S D	@L A	@S F	Cin	Atl	Hou
	@Cin	Atl	@L A	Hou	@S F	S D	9/14	@L A	@S F	@S D	Atl	Hou	Cin
6/22	S F	@Hou	Cin	@S D	L A	@Atl		@S F	Hou	@Cin	S D	@L A	Atl
	Cin	@Atl	L A	@Hou	S F	@S D	9/21	Cin	@Atl	L A	@Hou	@S F	S D
6/29	@Hou	S D	Atl	@S F	@Cin	L A		S F	@Hou	Cin	@S D	L A	@Atl
	@S F	Hou	@Cin	S D	@L A	Atl	9/28	@Cin	Atl	@L A	Hou	S F	@S D
7/6	Hou	@S D	@Atl	S F	Cin	@L A							

Away games denoted by @.

Summary

In this article, the schedule-maker for the National and American Base-ball Leagues describes the process he employs. First, the required features and desirable characteristics for a schedule are outlined.   Then, the con-cept of a home-and-away pattern is introduced and the algorithm for con-structing schedules from such a pattern is explained. The final parts of the article are a comparison between the 1975 and 1969 schedules to illus-trate the results obtained and a brief look forward.

THE VALUATION OF A BASEBALL PLAYER

by

Carl R. Mitchell and Allen J. Michel

Each year major-league sports-franchise owners negotiate contracts with their players. To develop an effective negotiating strategy, the owners must calculate the maximum value of a compensation package (salary, bonus, retirement plan, etc.), beyond which the marginal cost of the player exceeds his marginal value to the team. The focus of this paper is on the development of a methodology to determine the present value of a player. The analysis below is restricted to the consideration of a star major-league pitcher; similar methodology is applicable to other players and/or other sports. Typically, market values will provide a guideline in compensation determination; however, market values become operationally difficult to use when determining the value of a star player.

A simulation approach is utilized. Such simulations have frequently been applied in a business context [Carter, 1972; Economos, 1968; Hertz, 1964], but rarely in an athletic environment. From the viewpoint of management, the problem can be considered similar to a capital expenditure. In other words, a net-present-value (NPV) or internal-rate-of-return (IRR) analysis can be utilized. However, due to the stochastic nature of the variables, a simulation approach is undertaken. A model is developed and used with actual data supplied by a major-league baseball team to determine how much that team would have paid for "Catfish" Hunter who signed with the New York Yankees in the winter of 1975.

Table I lists those probability distributions which must be generated as inputs to the model. These distributions are based primarily on management's subjective estimates of both Hunter's and the team's performance. Bayesian statistics can be useful in combining these estimates of performance with a likelihood function generated from past data [Raiffa, 1970]. Table II illustrates a sample of actual data generated by a major-league team.

Table I

Required Input Distributions

Career Lifetime
Number of Wins
Incremental Home Attendance when Hunter Pitches
Number of Home Games in which Hunter Pitches
Incremental Attendance when Hunter does not Pitch
Probability of Winning Division Title
Probability of Winning Pennant
Probability that Division Title is due to Hunter
Probability that Pennant is due to Hunter
Attendance at Playoff Games, World Series
Number of Playoff Games, Series Games
Incremental Salary Increase of Other Players due to Hunter

Table II

Sample Data Describing Input Distributions

Probability that Hunter's Career Lifetime Exceeds:

1 year	2 years	3 years	4 years	5 years
98%	95%	85%	80%	75%

Probability that the Number of Wins in Year i Exceeds:

	0	5	10	15	20	25	30
(i=1)	100%	100%	99%	85%	50%	25%	1%
(i=2)	100%	95%	90%	80%	45%	20%	1%
.							
.							
.							
(i=5)	100%	90%	80%	65%	30%	5%	0

Number of Hunter Wins   Probability that Extra Attendance at Hunter Games Exceeds:

	0	2500	5000	7500	10,000	12,500	15,000	17,500
10	100%	95%	60%	20%	5%	0	0	0
20	100%	100%	100%	80%	50%	25%	10%	0

Together with the introduction of prices (i.e. price per seat), the previous distributions can be used to generate a frequency distribution of the present value of Hunter's worth. The simulation model is initiated by generating a simulated playing life for Hunter. Then the model is used to calculate Hunter's yearly revenue contribution during the regular season, the playoffs, and the World Series. The latter two contributions are dependent on whether winning the Division Title or winning the Pennant is attributable to Hunter's performance. The present values of the revenue generated each year by Hunter are then accumulated. From this total, the present value of the incremental increases in other player's salaries due to Hunter's salary is subtracted. The simulation is run many times and a frequency distribution of NPV or IRR is generated. Employing this distribution, management can select a contract price that is consistent with its own aversion to risk.

In addition to generating present values utilizing the above methodology, a sensitivity analysis of the results can be performed. The present value of the player may be relatively sensitive to some of the inputs while insensitive to others. Those to which it is most sensitive should in general be forecasted most accurately. On the other hand, if the PV is inelastic with respect to an input, a simple best guess will be a sufficient forecast. To approximate those variables to which the NPV is most sensitive, sensitivity values (S.V.) are calculated [Michel and Ostertag, 1974]:

$$S.V. = \Delta PV / \Delta\% \text{ input}$$

The S.V.'s are obtained by generating the present value of Hunter's contribution using the inputs described previously. The PV is then recalculated after shifting the expected value of one of the inputs $\Delta\%$. The change in present value, ΔPV, divided by the percentage change in the input, $\Delta\%$ input, is then defined as the sensitivity value of that variable.

Performing this calculation on the team's initial estimates of each variable allows the team to obtain a relative sensitivity ranking which reveals those critical variables which significantly impact the outcome of the simulation. Insensitive factors are left as originally estimated, while the more sensitive factors may be revised.

The probability distributions required for the simulation were generated by a major-league team to calculate the value to that team of acquiring Hunter. To make the analysis comparable to the actual situation, we asked that a five-year time horizon be used, since Hunter's contract runs for five years. The following table indicates the probability distribution of present values which was generated.

Table III
Present Value of Catfish Hunter

Probability	That the Present Value is Less Than:
0%	-425,000
10%	800,000
20%	1,300,000
30%	1,715,000
40%	2,085,000
50%	2,415,000
60%	2,710,000
70%	3,055,000
80%	3,480,000
90%	4,110,000
100%	6,968,000

The expected value of the distribution is 2,415,000, with a standard deviation of 1,287,500. These values, together with the others generated, enable the team management to use its risk-return trade-off to determine Hunter's compensation.

Because of the relatively subjective nature of the variables, it is important to calculate their sensitivity. Table IV lists the sensitivity of some of the key variables used in the analysis.

Table IV indicates that the sensitivity of the simulation to lifetime is the largest. Its forecast therefore is the most critical in using the analysis. The forecast of the number of wins each year is also critical, with the near-term values (i.e., 1975) being most important.

Notice that the sensitivity of the attendance is greatest when Hunter wins 20 games and least when he wins 30. This result is caused by the fact that it is much more probable that Hunter wins 20 games than 30.

Table IV
Sensitivity Analysis of Predicted Variables

VARIABLE	SENSITIVITY VALUE	RELATIVE SENSITIVITY VALUE	% CHANGE IN P.V. / % CHANGE IN INPUT
Lifetime	$13,843	1.0	0.57
# Wins (yr=1977)	9,509	0.70	0.40
(yr=1976)	9,662	0.70	0.40
(yr=1975)	10,792	0.79	0.45
Attendance at Games in which Hunter Pitches			
(wins=20)	4,703	0.33	0.19
(wins=25)	3,464	0.25	0.14
(wins=30)	222	0.00	0.00
Expected Salary Increase For Other Players	4,971	0.37	0.21
Attendance at Games in Which Hunter Does not Pitch			
(wins=10)	2,743	0.19	0.11
(wins=20)	3,823	0.28	0.16
(wins=25)	2,892	0.21	0.12
Play-Off Attendance	830	0.06	0.04
Probable Contribution to Winning Pennant			
(wins=20)	2,940	0.21	0.12
(wins=25)	1,583	0.11	0.06
Probability of Winning Division	2,422	0.18	0.10

Several simplifying assumptions have been made. It is assumed that all revenues generated at home games are kept by the home team. The visiting team actually receives a certain percentage of attendance revenues, but since the team plays half its season at home and half on the road, the assumption is probably not critical. Revenues other than attendance (e.g., vendor revenues) and the tax consequences of the treatment of a player as an asset by the team are not considered. In addition, termination value is not included. If Hunter is expected to retire he will have little or no value, but if he is either traded or remains with the team, there will be a positive termination value at the end of 5 years. Such considerations could be included without change in the basic methodology presented in this paper.

"BACK-TO-THE-WALL EFFECT": 1976 PERSPECTIVE

by

William Simon

Based on articles in Science*, 1971, and The Wall Street Journal, 1973

Of the thirty-one World Series played since the end of World War II, eighteen have lasted seven games. For equally matched teams the probability of a series lasting seven games is

$$P = C_3^6 \ (\frac{1}{2})^6 = 0.3125$$

where C is the binomial coefficient. The probability of eighteen or more out of thirty-one series lasting seven games is given by

$$\sum_{n=18}^{31} C_n^{31} \ p^n (1-p)^{31-n} = 0.0018$$

For unequally matched teams, the probabilities are lower. There are, of course, theoretical difficulties associated with the statistical interpretation of unusual events defined after the fact. If one examines enough records of enough sequences, a certain number of apparently bizarre results is bound to show up. Most of these will not be significant. For example, one might note that in the last 150 years, every president of the United States elected in a year ending in zero has died in office, and no other president has. In spite of the low a priori probability of this occurring, no one would seriously suggest that this pointed to a physical or sociological law.

To some degree one feels that the significance of the occurrence of improbable sequences is related to the level of ingenuity one needs to define the sequence. The seven-game phenomenon is straightforward and could easily be imagined by a person who did not know exactly what the World Series was, provided he had a sufficiently healthy level of cynicism toward the activities of his fellow man. But cynics should note that the TV contract (the principal source of funds) calls for a fixed sum, independent of the length of the series, and that the players' compensation is also invariant with the number of games.

Table I. Expectation and occurrence of the number of games in the World Series from 1905 to 1975**

No. of Games	Expectation p	Actual Occurrence 1905-1944	1945-1975
4	.125	6/37 = .162	4/31 = .129
5	.250	14/37 = .378	4/31 = .129
6	.313	9/37 = .243	5/31 = .161
7	.312	8/37 = .216	18/31 = .580

**The years 1919, 1920, and 1921 are excluded because the series consisted of the best five out of nine games.

---

*Parts of this article reprinted by permission. Copyright 1971 by the American Association for the Advancement of Sciences.

Upon further examination, the bizarre outcomes of the World Series trace to the result of the sixth game. Of the thirty-one World Series since the end of World War II, twenty-two have gone to the sixth game, and of these in only four cases has the team that was ahead going into the sixth game won the sixth game. If the teams are evenly matched, the probability of the leading team winning less than five times is

$$\sum_{m=18}^{22} C_m^{22} \left(\frac{1}{2}\right)^{22} = 0.0022$$

The theory has been advanced that there is a "back-to-the-wall" effect operating, which tends to favor the trailing team in the sixth game. This psychological explanation, however, is refuted by the fact that prior to World War II the sixth-game effect is not detectable; the lengths of World Series agree moderately well with theoretical predictions, with some bias toward shorter series, which is probably due to team imbalance. Furthermore, the obvious sources of biases which would tend to favor longer series, such as the home-park/visitor's-park phenomenon, have been tested and do not appear significant. The effect seems clearly associated with the sixth game, and with post-World War II baseball.

This back-to-the-wall effect was first pointed out in an article in Science, November 11, 1971, at which time fifteen out of twenty-seven post-World War II World Series had lasted seven games. The author was struck by the number of responses he received from readers who suggested underworld influences fixing the outcome of the World Series, or at least the length of the World Series. Two years later the article was up-dated, at which time the two intervening World Series had both lasted seven games. This time, which was during the Watergate crisis, not a single letter mentioned underworld influence; instead, several cited the term used to account for the famous eighteen-minute gap on the White House tapes: "sinister forces".

Editors' Note: this article was written before the 1976 series, which went four games. This weakens slightly the conclusion about seven-game series (now 18/32), but does not change the "sixth-game effect" (still 18/22). Dr. Simon tells us "I don't think even sinister forces could have stopped Cincinnati this year."

ADJUSTING BASEBALL STANDINGS FOR STRENGTH OF TEAMS PLAYED

by

John H. Smith

Based on an article in <u>American Statistician</u>, 1956

When a major-league team has played too many games with relatively weak opponents, its current standing is too large in comparison with teams whose opponents have been stronger on the average. Fairer comparisons can be made in terms of standings adjusted for strength of teams played. To this end, the probability p_{ij} that team i will win each game it is to play with team j is estimated for each pair of teams from results of games already played. In order that as much information as possible is used to estimate each p_{ij}, it is assumed, for simplicity, that each p_{ij} is of the form

$$p_{ij} = 0.5 + h_i - h_j \qquad (1)$$

where h_i and h_j are measures of strength of teams i and j, respectively.

The adjusted standing for team i is its probable standing if the schedule thus far had been proportional to that for the season; i.e. half again as many games with each team in its division as with each team in the other division.

Of any set of T_i games played by team i, the expected number of wins is

$$W_i = \sum_j p_{ij} T_{ij} \qquad (2)$$

where T_{ij} is the total number of games played by teams i and j, and $T_i = \sum_j T_{ij}$. Note that $T_{ij} = T_{ji}$ and $p_{ij} = 1 - p_{ji}$.

Equations (1) and (2) lead to a system of linear equations in the h's of which the typical equation can be written

$$\Sigma T_{ij} h_j - T_i h_i = T_i/2 - W_i \qquad (3)$$

from which values of the h's needed for estimating the p_{ij}'s can be found by successive approximations. The current standing of team i can be used as a first approximation for h_i, but the number of iterations is usually smaller when the better approximation

$$h'_i = p_i - (p_i - 0.5)/n \qquad (4)$$

is used, where n is the number of teams in the league (currently 12) and p_i is the "percentage" of team i in the current standings expressed as a fraction. Second approximations are then found by means of the formula

$$h''_i = (\Sigma T_{ij} h'_j + W_i - T_i/2)/T_i \qquad (5)$$

48

This may be repeated to find even better approximations, but one application should usually be sufficiently accurate, especially if the first approximation is found from (4). When these estimates are used for the h's in (1), the resulting p_{ij}'s have the desired properties. They lead to current standings when applied to games already played. When applied to all games scheduled for the season, they lead to standings adjusted for strength of teams played.

The adjustment-for-strength method is applicable only after the season has progressed sufficiently to provide the needed information. Strictly speaking, the system of equations (3) has a singular matrix and the minimum amount of information is that needed to make the rank of the matrix equal to n-1, so that differences of the form $h_i - h_j$ are determined uniquely. Actually, estimates of the p_{ij}'s are apt to be unreliable unless results of games played provide much more than the minimum information.

Adjusted standings should be interpreted with caution. We have assumed that probabilities of winning depend only on constant differences in team strengths. But there are surely changes in these probabilities. These changes may be short-run (as in choice of pitchers) or those of longer implications (such as effects of injuries and trades). One should not interpret adjusted standings as if all such neglected factors had been taken into account.

The adjustment-for-strength method is very flexible. It can be applied to current standings in other sports, such as basketball, where the number of games played is great enough to provide needed information. It can also be applied so as to answer questions other than those thus far suggested. For example, the estimated p_{ij}'s can be applied to a hypothetical schedule such as one in which the T_{ij}'s are equal regardless of division. Or standings may be adjusted a division at a time, ignoring interdivisional games.

To illustrate the adjustment-for-strength method, it is applied to National League games played through August 22, 1975, The T_{ij}'s which were listed in the body of the worksheet are omitted. Only the process of adjustment from the lower part of the worksheet is shown here, with the twelve teams listed in alphabetic order. The adjustments are small showing that variation in strength of opponents had little effect in this case. At a date very early in the season (e.g., mid May) adjustments would usually be larger and convergence to values of the h_i's slower.

ADJUSTMENT OF BASEBALL STANDINGS FOR STRENGTH OF TEAMS PLAYED
National League Games Played Team by Team through Aug. 22, 1975

	Atl	Chi	Cin	Hous	L.A.	Mont	N.Y.	Phil	Pitt	StL	SanD	SanF
Total	128	128	126	130	126	124	125	125	126	126	126	126
Won	57	60	83	48	67	53	65	69	71	67	57	61
Pct.	.445	.469	.659	.369	.532	.427	.520	.552	.563	.532	.452	.484
1st App.	.450	.472	.646	.380	.529	.433	.518	.548	.558	.529	.456	.485
W-T/2	-7	-4	20	-17	4	-9	2.5	6.5	8	4	-6	-2
2nd App.	.445	.472	.643	.379	.531	.435	.518	.549	.561	.529	.458	.482
Adj. pct.	.441	.466	.661	.368	.537	.425	.518	.552	.565	.530	.456	.482

Acknowledgment: I wish to thank Robert E. Machol for his help in the preparation and presentation of this article.

AN ANALYSIS OF BASEBALL AS A GAME OF CHANCE
BY THE MONTE CARLO METHOD

by

Earnshaw Cook[*]

Based on <u>Percentage Baseball</u>, 1964, and
<u>Percentage Baseball and the Computer</u>, 1972

Based upon simple theorems of probability, it was proposed that [Cook and Garner, 1964], except for the arrangement of the batting order, baseball tactics were misconceived. These misconceptions resulted in the improper use of sacrifice hits, attempted base steals, hit-and-run plays, and intentional walks. Further, baseball managers incorrectly evaluated player productivity, and used pitcher rotation and the platoon system ineffectively. These and other findings suggested that proper application of the principles of percentage baseball by any second-division club could, without change of personnel, elevate it to the first division and, in some instances, to the league championship.

<u>Percentage Baseball</u> also showed that the outcomes of annual World Series competition have primarily been subject less to the relative calibre of the teams than to the laws of chance. Subsequent analysis [Simon, 1971] has confirmed this observation.

Although it demonstrated that baseball is essentially a chance system, the 1962 volume could be criticized for at least two reasons:

(1) The theoretical and empirical equations of play appeared
     to possess <u>qualitative</u> reliability but offered little
     assurance of <u>quantitative</u> accuracy.

(2) It was too much to expect that mathematical equations of
     any sort could convince professional players, managers,
     and sportswriters that the conventional wisdom (which had
     been built on a century of experience) was wrong.

Therefore it was decided (in 1970) to undertake a different approach to the analytical appraisal of the game [Cook & Fink, 1972].

While baseball is an exceedingly complicated statistical process, it enjoys a relatively simple, repetitive pattern for the basic performances of nine offensive batsmen against the opposing pitcher supported by eight defensive players [Cook & Fink, 1972, pp. 44-57]. Thus, a sophisticated computer model of the game may be designed in which all possible interactions of play are precisely simulated. Chance determines each successive event; every play is selected by means of a random-number generator. The computer, in effect, functions only as an electromechanical scorekeeper; it summarizes all data for the 5,000 games of each simulation in less than five minutes, and reports the results in the same form in which they appear in the Official Baseball Guide. By this so-called 'Monte Carlo' method,

---

[*]Starting in 1962, Earnshaw Cook produced some of the earliest and best-known analyses of optimal strategies in sports. Although it is impossible to summarize here all of the content of his books, he has prepared this brief general summary of his methodology and conclusions.

it is possible to duplicate the average number of team runs scored for any season to within plus or minus one run, as well as to reproduce original team and player data to within the third decimal place.

The computer thus simulated on-the-field performance to achieve reliable mathematical models of major-league play (1951-1960) as well as those for three other systems included in the text: (1) the abortive strike-zone experiment of 1963, mercifully abandoned in 1968; (2) actual team play of the New York Yankees of 1970; and (3) of the Kansas City Royals of 1970. With the achievement of accurate simulations, two hypotheses were presented:

(1) Only by impeaching the design, construction, and/or operation of the model could its conclusions be logically denied. (Such indictment is most unlikely and none has been proposed to date by either professional or academic authorities.)

(2) If the integrity of the model is accepted, the quantitative positive or negative effects upon the scoring of runs, after introducing variations of strategy or performance, should provide reliable criteria for all features of play, as illustrated in Table I.

It may be observed that three _simultaneous_ variations producing 178 runs closely approach the sum (174 runs) of _separate_ introductions of the identical changes for successive simulations. It seems notable that the simulated increment of 73 runs for the substitution of four pinch-hitters for pitchers (#2c) is of the same order of magnitude (70) as that contributed by the Designated Hitter recently introduced in the American League (Table II).

A brief review of research findings includes the following [Cook & Fink, 1972, pp. 198-207].

(1) The upward progression of home runs per year may point to a similar variation in terms of the resilience of the baseball itself. The so-termed _power factor_ of the game (the quotient of total bases and total hits), which represents the average number of bases per hit, extends from 1.250 to 2.250, and offers an accurate index of the relative resilience of the baseball. Home-run production is a reliable linear function of the power factor and has varied progressively upward over the years from 13 home runs per team per season in 1905 to 147 HR per team in 1961. Deliberate changes in the manufacture of the baseball have been repeatedly and unconvincingly denied by major-league officials. It has been most unfortunate that these critical physical characteristics of the ball have apparently not received the attention and rigid specifications demanded for the golf ball by the United States and Professional Golf Associations.

(2) Elimination of the sacrifice bunt from team play normally achieves no reduction in double plays; use of the sacrifice bunt contributes an average loss of about 30 runs per season.

(3) The intentional walk _at usual frequencies_ increases opponent's scoring by about 3 runs per season with no increase in double plays.

Table I

Scoring Effects of Simultaneous versus Successive Introductions
of Variations of Play: Kansas City: 1970

	Runs Scored:		
	Total:	Basic:	Additional
(1) Simultaneous introduction of			
(a) 8 Most Productive Men			
(b) + Four Pinch Hitters			
(c) + Elimination of Sacrifice Bunt	775 R	601 R	+ 174 R
(2) (a) 8 Most Productive Men only:	673	601	+ 72 R
(b) + Elimination Sac. Bunt only:	706	673	+ 33 R
(c) + Four Pinch Hitters only:	779	706	+ 73 R
(3) Cumulative Total: 2a + 2b + 2c =			+ 178 R
(4) From #1, above (simultaneous introduction)			+ 174 R

(From Cook and Fink, 1972, Table 44; p. 145).

Table II

Official Data:	Runs Scored: 162 Games:
1971: American League Conventional Play:	623 R
1973: American League + Designated Hitter:	693 R

       Scoring Advantage: Four Pinch Hitters: + 70 Runs
       [Compare with 2c above]

(4)   The stolen base, at an average success rate of .55, appears to
exert a negative 3-run effect upon scoring for the 2-out situation. However, for the no-out situation and an average success
rate of over 70 percent, significant advantages of up to 60
runs per season exist.

(5)   Despite reductions in double plays, the adverse effects of the
mandatory stolen base on a missed strike operate against the
apparent value of the hit-and-run play to decrease scoring in
all situations by about 3 runs.

(6)   The general influence of intuitional platooning is variable, but
failure to identify and play the eight most productive players for
90% of the time can reduce annual team scoring by as much as 75
runs.

(7)   Optimization of batting orders appears to follow the traditional
format of leading off with two players of high on-base potential,
followed by three batsmen arranged in ascending order of their
slugging averages. The most effective line-ups may be determined only by a series of computer simulations.

While the usual RBI statistics for 162 games are too erratic for reliability, they become remarkably consistent and revealing in the duplicated simulations of 5,000 games. A very uniform, constant difference of 28-30 runs exists between team runs scored and batted-in, regardless of the batting order used (Table III). However, various batting orders have different run-production capabilities. This involves the interactions of the performances of individual players, as illustrated in the table. Players are listed from "A" thru "P", in descending order of their respective indices of scoring potential, i.e., expected runs contributed to the team. The conventional batting order (KC-4) produced 601 runs, or 21 more runs than the ascending order (KC-11) with the pitcher leading off, and 13 more runs than the descending order (KC-13).

Item KC-14 was processed to determine the hypothetical effects of batting the same player (T) in all nine positions. Since the only statistical difference existed in the line-up positions, this arrangement developed the probable positional influence of the batting order itself upon individual run producing potential:

Table III

Effect of Changing Composite Batting Order Positions

KC-4		KC-13		KC-11		KC-12		KC-14	
Conventional		Descending		Ascending		Scrambled		Team Comp Avg All Players	
Player	RBI	Player	RBI	Player	RBI	Player	RBI	Player	RBI
A	55.0	A	56.1	P	24.8	D	67.5	T	58.4
E	52.8	B	82.4	H	43.5	F	72.2	T	60.7
C	76.1	C	70.1	G	44.8	H	53.6	T	62.4
F	86.3	D	72.8	F	65.6	B	102.1	T	70.0
B	102.5	E	67.7	E	58.1	P	24.5	T	65.6
D	66.2	F	72.9	D	66.3	A	58.1	T	62.8
G	56.5	G	60.9	C	79.7	G	51.0	T	61.3
H	50.3	H	50.7	B	96.5	E	52.9	T	59.9
P	24.5	P	24.5	A	70.3	C	67.9	T	57.6
Run	600.8		587.1		579.9		579.8		588.9
RBI	570.2		558.1		551.1		549.8		558.7
d.f.	30.6		29.6		28.8		30.0		30.1

Positions: Difference from Average RBI (62.1)
1, 9:   -3.7 to -4.5 RBI
2, 8:   -1.4   -2.2
3, 7:   -0.3   -0.8
6:   +0.7
5:   +3.5
4:   +7.9

Thus, moving the same average player from positions #1 or #9 to the most favorable fourth position would result in an expected increase (for that player) of about 12 RBIs for the season. Other data confirm that this effect is primarily a result of the increased number of runners on the bases when the #4 batsman is at the plate.

"Playing the percentages", annually proclaimed by field managers as a basic feature of the futile attempts of at least eleven of them in each league to win a championship, fundamentally has to be an exercise in sheer ignorance and a display of their contempt for the scientific method. Knowing the correct odds for what is going to happen in a random system most of the time is beyond the mental and physical capacities of the most astute natural or athletic scientists without the constant use of meticulously programmed, digital computers. Baseball at its best is a game of speed, grace, daring, and skill. While the probability of a base hit or a strike-out may precisely be determined, the real charm of the game is that no one will ever know "when"! Indeed every field manager might profit by giving heed to Yogi Berra's immortal question to the greatest of all managers, Casey Stengel: "Now, Coach, how can you think and bat at the same time?"

The remarkable Monte Carlo Method should and probably will not be denied an eventual contribution to the National Game of America.

THE DISTRIBUTION OF RUNS IN THE GAME OF BASEBALL

by

D. A. D'Esopo and B. Lefkowitz[*]

The mathematical model of the national pastime presented here differs
in a number of ways from the game familiar to millions of fans. First,
players are assumed to be equal in batting ability, so that one set of probabilities describes the expected performance of all batters. Also, these
probabilities are assumed to remain fixed during play; clutch hitting is
not accounted for in the model. Second, it is assumed that all activity on
the diamond must involve the batter, i.e., plays such as stolen base, wild
pitch, balk, passed ball, and pick-off are excluded. Third, it is assumed
that base runners advance only when the batter does not make out; therefore,
bunts and sacrifice flies are treated as ordinary outs. Fourth, it is
assumed that non-outs are of just five types, called, for the sake of clarity, walk, single, double, triple, and home run, and that each of these particular plays always produces one effect. Consequently, hit-and-run plays
and heads-up base running are unknown in the model. The plays available
in the model game and their assumed effects are enumerated in Table I.

It must be admitted that the model game probably would be a dull one
to watch: No derring-do on the base paths; no fielding gems; no managerial
blunders; only six events altogether (including out). Despite the simpli-
fications, however, the model game retains important elements underlying
run production in the real game. The results of the analysis support the
belief that the plays which are excluded from the present formulation,
though important to the outcome of individual games, make a minor contribution to the distribution of runs scored.

Table I

Effects Produced by Different Baseball Plays

Play	
Out	Base runners do not advance.
Walk	Batter takes first. All base runners advance one base if forced to do so.
Single	Batter takes first. Base runner on first takes second. All other base runners score.
Double	Batter takes second. Base runner on first takes third. All other base runners score.
Triple	Batter takes third. All base runners score.
Home Run	Batter scores. All base runners score.

[*]The present paper is an edited version of one that was presented at the
1960 Annual Meeting of the American Statistical Association. ASA subsequently accepted it for publication, but it was never printed. This version incorporates changes suggested by Robert E. Machol as well as those
of the referees whose work so long ago can finally be acknowledged. The
authors also wish to thank the Sports Department of the San Francisco
Chronicle for making available the statistics used in this paper.

The Distribution of Hits in a Half-Inning. Let p be the probability of
a non-out or, as it will be subsequently called, p-hit. Then π_n, the probability of obtaining n p-hits in a half-inning, is the number of ways n
p-hits and three outs can be combined (given that a combination ends in an
out) times the probability of n hits and three outs in some specific order.

$$\pi_n = \binom{n+2}{2} p^n (1-p)^3 \qquad (1)$$

That is, the number of p-hits in a half-inning follows a negative binomial
distribution.

The Distribution of Runs in a Half-Inning. If all p-hits were home
runs, the probability of n runs in a half-inning, ψ_n, would be precisely
the probability of n p-hits, π_n. If all p-hits were singles,

$$\psi_0 = \pi_0 + \pi_1 + \pi_2$$

$$\psi_n = \pi_{n+2}, \; n > 0$$

That is, it would take three p-hits to score the first run, and each additional p-hit would produce another run. Similar expressions can be worked
out for the other p-hits. In general, however, knowing the number of p-
hits in a half-inning is not sufficient to predict the number of runs that
will be scored. The number of runners who cross the plate depends on both
the kinds of p-hits produced and the sequence in which they occur. For
example, a double followed by a single will score a run while (according
to the assumptions in Table I) a single followed by a double will not.

These considerations suggest a procedure for computing the distribution of runs scored in a half-inning. First, enumerate all finite p-hit
sequences and, using the rules in Table I, classify each by the number of
runs produced. (Note that it is not possible to score n runs with sequences
of less than n, nor more than n+3, p-hits.) Next, compute the probability
of occurrence for each sequence. Finally, sum these probabilities for
sequences which produce the same number of runs. The totals are the ψ_n.

The procedure can be illustrated for the probability ψ_0, that no runs
are scored in a half-inning. Table II presents a complete enumeration of
sequences that score no runs grouped by the number of p-hits in the sequence.
The probability of obtaining a specific sequence of p-hits is the product
of: (1) The probability of obtaining the number of p-hits in the sequence
π_n and (2) the conditional probabilities of those p-hits. For example, the
probability of obtaining the sequence single-double-walk in a half-inning
in that order (with perhaps some of the outs intervening) is

$$\alpha_1 \alpha_2 \alpha_0 \pi_3,$$

where α_i, i=0,...,4, is the conditional probability that a p-hit is a walk,

single, double, triple, and home run respectively ($\Sigma \alpha_i = 1$).

When all sequences in Table II have been evaluated and summed, the
result is

$$\psi_0 = A_0 \pi_0 + B_0 \pi_1 + C_0 \pi_2 + D_0 \pi_3 \tag{2}$$

where

$$A_0 = 1$$

$$B_0 = \alpha_0 + \alpha_1 + \alpha_2 + \alpha_3 = 1 - \alpha_4$$

$$C_0 = \alpha_0(\alpha_0 + \alpha_1 + \alpha_2) + \alpha_1(\alpha_0 + \alpha_1 + \alpha_2) + \alpha_2\alpha_0 + \alpha_3\alpha_0$$

$$D_0 = C_0\alpha_0$$

In general,

$$\psi_n = A_n \pi_n + B_n \pi_{n+1} + C_n \pi_{n+2} + D_n \pi_{n+3} \tag{3}$$

where

A_n = probability n runs are scored with n p-hits

B_n = probability n runs are scored with n+1 p-hits

C_n = probability n runs are scored with n+2 p-hits

D_n = probability n runs are scored with n+3 p-hits

The same steps--enumeration, computation and summation--could be repeated to get ψ_1, ψ_2, ...

Fortunately, this does not have to be done. It is a curious fact that it is only necessary to enumerate the nonscoring sequences. To go from nonscoring sequences to one-run sequences, observe the following principles:

1. Any one-run sequence is the result of placing a p-hit in front of some nonscoring sequence containing one less p-hit. For, suppose S_1 is a one-run sequence, then it must have at least one p-hit. Dropping the first p-hit of S_1 produces a nonscoring sequence (call it S_0). The sequence S_0 is nonscoring because the first p-hit in S_1 must have been the run that scored; remove it, and the run no longer scores.

2. A p-hit placed in front of a nonscoring sequence produces a new sequence which scores either (a) no runs, or (b) one run.

Table II

P-Hit Sequences that Score No Runs

Number of p-hits	Sequences
0	all
1	walk; single; double; triple
2	walk-walk; walk-single; walk-double; single-walk; single-single; single-double; double-walk; triple-walk
3	walk-walk-walk; walk-single-walk; walk-double-walk; single-walk-walk; single-single-walk; single-double-walk; double-walk-walk; triple-walk-walk

To illustrate the use of these principles, consider a nonscoring, one p-hit sequence S. When S is preceded in turn by each possible kind of p-hit, it becomes either a nonscoring, two p-hit sequence, or a one-run, two p-hit sequence. When this process is applied to every nonscoring, one p-hit sequence, it results in the generation of every nonscoring, two p-hit sequence and every one-run, two p-hit sequence, and nothing else. As a consequence of this partitioning.

$$B_0 = B_1 + C_0$$

By the same reasoning

$$A_0 = A_1 + B_0, \quad C_0 = C_1 + D_0, \quad D_0 = D_1$$

Solving for A_1, B_1, C_1, and D_1 gives

$$A_1 = A_0 - B_0, \quad B_1 = B_0 - C_0, \quad C_1 = C_0 - D_0, \quad D_1 = D_0 \qquad (4)$$

The generation of the coefficients A_n, B_n, C_n, and D_n for $n > 1$ is even easier. To obtain them observe that a p-hit placed in front of any run-scoring sequence must increase by one the number of runs scored. (The base runner produced by the additional p-hit will figuratively be "pushed" across the plate by the runner who scores after him.) For example, placing a walk in front of the one-run, two p-hit sequence "single-triple" turns it into a two-run, three p-hit sequence, "walk-single-triple". Considerations of this type yield: B_2 (probability of scoring two runs on three p-hits) = B_1 (probability of scoring one run on two p-hits); B_3 (probability of scoring three runs on four p-hits) = B_2. And in general: $A_1 = A_n$, $B_1 = B_n$, $C_1 = C_n$, $D_1 = D_n$, ($n \geq 1$). Consequently, (3) can be written

$$\psi_n = A_1 \pi_n + B_1 \pi_{n+1} + C_1 \pi_{n+2} + D_1 \pi_{n+3}, \quad n \geq 1 \qquad (5)$$

Combining (5), (4), and (2) gives the desired distribution of runs in a half-inning ψ_n. These equations amount to a description of ψ_n in terms of four independent parameters: p, B_0, C_0, and D_0 (recall that $A_0 = 1$). Further, this reduction does not depend strongly on the assumed effects shown in Table I. In fact, the expression of ψ_n in terms of four parameters is valid under very general assumptions, including some in which p-hits do not have a determinate effect on scoring. What is required is that scoring follows the principles given above.

The Average Number of Runs in a Half-Inning. The average number of runs in a half inning μ, is

$$\mu = \sum_{n=0}^{\infty} n\psi_n = \sum_{n=1}^{\infty} n\psi_n \qquad (6)$$

which expanded gives

$$\mu = A_1 \sum_{1}^{\infty} n\pi_n + B_1 \sum_{1}^{\infty} n\pi_{n+1} + C_1 \sum_{1}^{\infty} n\pi_{n+2} + D_1 \sum_{1}^{\infty} n\pi_{n+3} \qquad (7)$$

Substituting the expressions in (4) and collecting terms

$$\mu = A_0 \sum_1^\infty n\pi_n + B_0 \sum_1^\infty (n\pi_{n+1} - n\pi_n) + C_0 \sum_1^\infty (n\pi_{n+2} - n\pi_{n+1})$$

$$+ D_0 \sum_1^\infty (n\pi_{n+3} - n\pi_{n+2}) \tag{8}$$

Since

$$\sum_{n=1}^\infty (n\pi_{n+j} - n\pi_{n+j-1}) = \sum_{n=1}^\infty \{ (n-1)\pi_{n+j-1} - n\pi_{n+j-1} \} = -\sum_{n=1}^\infty \pi_{n+j-1} = -\sum_{n=j}^\infty \pi_n \tag{9}$$

(8) may be simplified to

$$\mu = \sum_1^\infty n\pi_n - B_0 \sum_1^\infty \pi_n - C_0 \sum_2^\infty \pi_n - D_0 \sum_3^\infty \pi_n \tag{10}$$

The first term on the right-hand side of (10) represents the expected number of p-hits in a half-inning; its value is $3p/(1-p)$. Consequently, the remaining terms, which can be evaluated using (1), represent the expected number of men left on base.

Problems of Estimation. Unlike other, less important, human undertakings, baseball is a well documented activity. Summary statistics are available on all aspects of the game: Fielding, pitching, batting. Unfortunately, tabulated baseball data are not precisely in the form required to evaluate performance in our model game. For one thing, the underlying probability p, of a p-hit, is not the same as its nearest real-game counterpart--the batting average. The latter does not credit batters for walks, hit by pitch, or errors, nor does it penalize batters for sacrifices and double plays. Our approach is to determine $\hat{p}$, the estimate of p, from the expression

$$\hat{p} = \frac{\text{Hits} + \text{Walks} + \text{Hit by Pitch} + \text{Errors} - \text{Double Plays}}{\text{At Bats} + \text{Walks} + \text{Hit by Pitch} + \text{Sacrifices}} \tag{11}$$

Note that double plays are treated as negative p-hits. All quantities in (11) are available in published form. [Official Baseball Guide, 1960].

In estimating α_j, it is apparent that hit by pitch (HBP) should be treated as a walk. We were less certain about errors and double plays, and allocated them, somewhat arbitrarily, to $\hat{\alpha}_0$ and $\hat{\alpha}_1$ as shown below. The estimates are defined as follows: Let P be the total number of p-hits, i.e.,

$$P = \text{Hits} + \text{Walks} + \text{HBP} + \text{Errors} - \text{Double Plays}$$

Then,

$$\hat{\alpha}_0 = \frac{1}{p} \left\{ \text{Walks} + \text{HBP} + \frac{\text{Walks} + \text{HBP}}{\text{Walks} + \text{HBP} + \text{Singles}} \ (\text{Errors} - \text{Double Plays}) \right\}$$

$$\hat{\alpha}_1 = \frac{1}{p} \left\{ \text{Singles} + \frac{\text{Singles}}{\text{Walks} + \text{HBP} + \text{Singles}} \ (\text{Errors} - \text{Double Plays}) \right\}$$

$$\hat{\alpha}_2 = \frac{1}{p} \ (\text{Doubles}) \tag{12}$$

$$\hat{\alpha}_3 = \frac{1}{p} \ (\text{Triples})$$

$$\hat{\alpha}_4 = \frac{1}{p} \ (\text{Home Runs})$$

Computed Results. Table III contains summary information for all 1959 National League games needed to compute $\hat{p}$ and the $\hat{\alpha}_i$. Using this data, formulas (11) and (12) yield the following values:

$$p = .320123, \ \alpha_0 = .276074, \ \alpha_1 = .508303, \ \alpha_2 = .117865,$$

$$\alpha_3 = .021358, \ \alpha_4 = .076401 \tag{13}$$

Table IV compares the expected distribution of runs scored in a half-inning as computed using formulas (2) and (4) and the above estimates of p and α_i, with two observed sample distributions. The first is derived from 100 National League games in 1959, excluding the last half of the ninth inning (the choice of sample was motivated by the authors' interest in the San Francisco Giants, who make up half the observations). The second is derived from Lindsey's data on completed half-innings in 1000 National and American League games played in 1959 (see Lindsey's article p. 9).

The agreement is quite good, surprisingly so because neither sample corresponds to the data used to estimate the model's parameters. The large number of four-run innings in our sample appears to be a statistical fluke. The consistent differences in frequencies of zero-run and one-run innings probably can be attributed to the model's assumption that all players are equal. In actuality, the heavy hitters at the top of the lineup probably produce an excess of one- and two-run rallies that are snuffed out by the weaker hitters at the tail-end of the order.

Applications and Extensions. Lindsey and other authors have discussed limitations of the batting average--the ratio of hits to at bats--as a measure of effectiveness. Another measure of effectiveness, the slugging

Table III

Selected 1959 National League
Baseball Statistics [Baseball Guide, 1960]

At Bats	Hits	Hit By Pitch	Walks	Singles	Doubles	Triples	Home Runs	Errors	Double Plays	Sacrifices
42330	11015	232	3974	7744	1788	324	1159	1113	1164	852

Table IV

Observed and Expected Distributions
of Runs Scored in a Half-Inning

Runs Scored	D'Esopo-Lefkowitz	Lindsey	Expected
0	.744	.730	.758
1	.129	.146	.123
2	.068	.070	.065
3	.029	.029	.031
4	.021	.014	.014
5	.007	.007	.006
≥ 6	.003	.004	.004
	.489	.488	.456

average, weights hits according to a particular value scale: One for single, two for double, three for triple, four for homer. Neither measure seems to comprehend a model of run production, else events such as walk and double play would figure in the calculation. Because run production is the primary objective of the team at bat, it would be useful to be able to judge individual batting performance in terms of that goal. We propose to do this by calculating how well a team of nine identical batters would do, i.e., the average number of runs it would score in an inning. This is nothing more nor less than μ in (10). Unfortunately, there are inadequate figures on how many times an individual batter got on base via fielding errors, the number of double plays he hit into, and how many times he was hit by a pitch. Consequently, (11) and (12) must be truncated.

With this in mind, it is possible to compare the performance of the 15 leading National League batters in 1959 on the basis of the traditional measures--batting average and slugging average--and the proposed new measure called here the scoring index. Table V shows this comparison (the slugging average and scoring index ranking are relative to the top 15 batters as determined by the batting average). The scoring index and slugging average rankings are similar; changes in position are due primarily to frequency of walks. Cunningham of St. Louis led all others with 88 bases on balls, and this contributed greatly to his first-place position in the scoring index rankings.

The preceding analysis leaned heavily on (1) the reduction of available baseball plays to six deterministic events, and (2) the implied independence of the repeated application of these events. Similar, but more complicated, synthetic games than the one described here have been analyzed by Howard, Bellman, and Trueman elsewhere in this book.

Table V

Performance of Leading Fifteen Hitters in the National League, 1959

Name	Batting Average	Rank	Slugging Average	Rank	Scoring Index	Rank
Aaron, Milw.	.355	1	.636	1	1.0732	2
Cunningham, St.L.	.345	2	.478	10	1.0755	1
Cepeda, S.F.	.317	3	.522	6	.7431	10
Pinson, Cincin.	.316	4	.509	7	.7696	9
Mays, S.F.	.313	5	.583	5	.9101	5
Temple, Cincin.	.311	6	.430	13	.7150	11
Robinson, Cincin.	.311	7	.583	4	.9406	4
Boyer, St.L.	.309	8	.508	8	.8254	8
Mathews, Milw.	.306	9	.593	3	.9598	3
Banks, Chi.	.304	10	.596	2	.8970	6
Moon, L.A.	.302	11	.495	9	.8301	7
White, St.L.	.302	12	.470	11	.6404	13
Hoak, Pitts.	.294	13	.399	15	.6346	14
Bell, Cincin.	.293	14	.449	12	.5716	15
Logan, Milw.	.291	15	.411	14	.6443	12

Summary. This paper presents a greatly simplified version of baseball; more specifically, it presents a model of the way in which runs are scored in a typical half-inning, ignoring "unusual" plays in explaining average run production. In this model, runs are scored as a result of success at the plate alone, and success is measured by the probability that a batter does not make an out. While the probability of a team amassing n non-outs in a half-inning is readily derived, this statistic is not sufficient to determine the amount of scoring, which depends on the different kinds of non-outs and their sequence of occurrence. It is assumed that there are five types of non-outs, each with its own effect on run production. A surprisingly simple way of combining non-outs into the distribution of runs is developed. The expected number of runs in a half-inning is also derived. The paper examines some problems in estimating the model's parameters from available baseball statistics, and then compares the observed and theoretical values for the distribution of runs scored and the expected number of runs per half-inning. The latter statistic is used to compare the batting effectiveness of 15 leading batsmen in the National League in 1959.

MONTE CARLO ANALYSIS OF BASEBALL BATTING ORDER

by

R. Allan Freeze
Based on an article in Operations Research, 1974

In 1964 Earnshaw Cook published the first edition of his massive statistical study of baseball. To the delight of many scientists and mathematicians (but to the bored indifference of most professional baseball men)
he used statistical analysis to attack many of the traditional strategies
of the game. The results of his inquiry received wide press coverage.

One of his conclusions [Cook, 1964] was that the batting lineup should
be arranged in decreasing order of productivity, with the teams' best batters
in the first positions rather than in their traditional cleanup positions.
He estimated a small but measurable improvement in team performance (on
the order of 11 runs per season) from this change in strategy. The purpose
of this brief paper is to report the results of a more definitive study of
the question of batting order. The details can be found in [Freeze, 1974].
The conclusions are based on a large number of Monte Carlo simulations
using a rather sophisticated baseball simulator. This approach provides
realistic consideration of the interactions between pitching and hitting,
between the various individuals on a team, and between the groups of individuals that make up competing teams.

Simulations were carried out with a programmed embodiment of the main
features of the Sports Illustrated Baseball Game, a commercially available
game that allows a sophisticated and accurate recreation of major-league
games. The performances of the players in the game (both batters and pitchers)
reflect their actual performances as evidenced by their past records.

The program was written in FORTRAN IV and simulations were carried out
on an IBM 360/91. The program required 80K storage. A set of 10,000 games
can be simulated in less than three minutes of computer time on this machine.

Output from the simulations can be specified in several forms: as
inning-by-inning, player-by-player game summaries; as simple line-score
summaries; or as overall summaries of wins, runs, hits, and home runs for
each team after a set of games. In all cases, the final output provides a
summary for each player of each team, listing times at the plate, walks,
at bats, hits, home runs, and batting average.

The simulated results were checked in a multitude of ways against real
statistical data. This was done at the individual level, at the team level,
on a game basis, and on a seasonal basis. In every test, the simulated data
showed good statistical correspondence with the real data.

To provide statistical fodder for the simulations, I chose two groups
of New York Yankee players, one representing nine of the batters and four
of the pitchers on the 1970 team (Team A), and the other representing a
composite lineup of all-time Yankee greats (Team B). A comparison of the
average values for Team A for each of several batting indices in comparison
with overall major-league averages for these indices (as provided by Cook)
shows that this team represents an average grouping of major-league players.
Team B, on the other hand, is of Hall-of-Fame caliber.

In order to arrange batting lineups in order of productivity, it is
necessary to define a single-parameter batting-performance index that can
be used as a measure of productivity. Cook suggests a parameter that he
calls the scoring index. He used the symbol DX for this parameter and defined it as:

$$DX = [P(1B) + P(E) + P(BB) + P(HP)] \cdot$$

$$[4P(HR) + 3P(3B) + 2P(2B) + P(1B)]$$

where P denotes a probability and E, BB, HP, 1B, 2B, 3B, and HR refer to
error, walk, hit-by-pitch, single, double, triple, and homerun, respectively.
Each batter has a DX-value that reflects his batting abilities.

Table I lists the various batting orders used in the study. The ATR
lineup is the traditional lineup for Team A. The ADX lineup has the players
of Team A ordered from the highest to the lowest DX-values. The ADXR lineup is the reverse of ADX. The BTR, BDX and BDXR lineups have the same
meaning for Team B. Of these, the TR and DX lineups are used to reach the
fundamental conclusions as to the relative worths of traditional and ordered
lineups, while the DXR lineups are used to determine a measure of the
largest possible influence that batting order can provide. Of the subsequent
lineups listed on Table I, the HDD lineups are used to locate the best position (first, fourth or ninth) for a consistent, powerful batter in a weak
lineup; the EDD lineups for a consistent, powerful batter in a relatively
strong lineup; and the HB lineups for a consistent, but not powerful, batter
in a weak lineup. Monte Carlo simulations were carried out that pit various
pairings of the lineups listed in Table I against one another (and for check
purposes, against themselves). The fact that many of these lineups and
pairings would be physically impossible might provide consternation in the
real world, but not, of course, in the world of simulation.

Conclusions to the batting-order question are drawn on the basis of
games won. Table II lists the numbers of games won by the winning lineup
n_1 in each of 22 N-game simulations where N = 10,000. The number of games
won by the losing lineup is of course $N-n_1$. The lineup designations are
those outlined in Table I. On the basis of statistical theory, one would
expect n_1 to possess a binomial distribution; and, since the number of games
N is large, one can use the normal approximation to the binomial to draw
conclusions. Thus, n_1 should be normally distributed with mean $\mu = Np$ and
standard deviation $\sigma = Np(1-p)$, p being the probability of a win. If the
two opposing lineups are identical, as is the case for runs 1, 2, 3, 7, 8
and 9 in Table II, then p = $1/2$, $\mu = 5000$, and $\sigma = 50$. It can be noted
/that, on these six runs, the maximum value of $n_1-\mu$ is 1.1σ, about as
expected.

In view of the results of runs 4-6, 10-12, and 18-22, where the opposing
lineups only involve variations in the ordering of the same set of players,
it is strikingly clear that batting order exerts only a small influence on
the outcomes of baseball games. Let us set up the hypothesis that the outcome of baseball games is independent of batting order for each of the 11
runs under consideration. We can consider the hypothesis accepted at the
95% confidence level if $n_1 -\mu < 1.96\sigma = 98$. In Table II the hypothesis
is accepted for runs 4, 6, 10, 18, 21, and 22 and rejected for runs 5, 11,
12, 19 and 20.

Turning to runs 13 through 17 in Table II, where various lineups of Teams A and B face each other, we would no longer expect $p = 1/2$. An estimate of its new value ($p = 0.8214$) can be obtained from run 13 where the traditional lineups oppose each other. For runs 13-17, $\mu' = 8214$ and $\sigma' = 38.4$. The hypothesis is now accepted when $|n_1 - \mu| \leqslant 75$. In Table II the hypothesis is accepted for runs 13, 14 and 16 and rejected for runs 15 and 17.

If we now examine the cases for which the hypothesis has been rejected, it is clear that we have proved statistical superiority for the traditional and DX-ordered lineups over the DX-reverse-ordered lineup, and for the positioning of strong hitters in the first or fourth positions rather than the ninth position of an otherwise weak lineup. These conclusions are hardly surprising. What may be surprising is the very small influence of what most baseball connoisseurs would consider to be a monumentally absurd strategy: the largest recorded difference between traditional and DX-reversed orders over 10,000 games is 130 (run 11). This comes out to just over 2 wins per 162-game season. The difference between putting Babe Ruth ninth or fourth in a team of Gene Michaels (run 19) is just over three wins per season!

On any individual set of 10,000 games between traditional and DX-ordered lineups (runs 4, 10, 14, and 16) it was not possible to reject the hypothesis that wins are independent of these batting orders. In each case, however, the traditional lineup provides superior performance (runs 14 and 16 must be compared with run 13 to reach this conclusion). If we group the four runs into a single sample with N = 40,000 it is possible to reject the hypothesis for the grouped data, and we can conclude that the traditional lineup is superior to the DX-ordered one. This conclusion apparently holds for both evenly matched teams and ill-matched ones.

The application of this type of simulator rests on a set of assumptions that tends to offend many baseball fans. It presumes a steady-state batting ability for each player, so that the probabilities for his various possible performances remain unchanged from at-bat to at-bat; thus, it denies the existence of a consistent clutch-hitting ability by an individual player (a phenomenon that may or may not exist), and the altering of hitting objectives from situation to situation (a phenomenon that certainly does exist). The liberties taken with the lineup in this study also contain an implicit assumption that a player's batting record is independent of his position in the batting order. This fails to take into account such pressures as those placed on cleanup hitters to hit the long ball. In short, then, the psychology and many of the subtleties are missing. The question of whether their absence significantly affects the statistical conclusions reached in this type of study is probably unanswerable.

Table I

Batting Orders Used in the Simulation

ATR	A	B	C	D	E	F	G	H	I
ADX	D	E	B	C	A	F	G	H	I
ADXR	I	H	G	F	A	C	B	E	D
BTR	AA	BB	CC	DD	EE	FF	GG	HH	II
BDX	DD	EE	BB	CC	FF	GG	AA	HH	II
BDXR	II	HH	AA	GG	FF	CC	BB	EE	DD
HDD1	DD	H	H	H	H	H	H	H	H
HDD4	H	H	H	DD	H	H	H	H	H
HDD9	H	H	H	H	H	H	H	H	DD
EDD1	DD	E	E	E	E	E	E	E	E
EDD4	E	E	E	DD	E	E	E	E	E
HB1	B	H	H	H	H	H	H	H	H
HB4	H	H	H	B	H	H	H	H	H

Teams and Scoring Indices

Team A. New York Yankees (1970): A - Clarke (.086), B - Cater (.117),
     C - Murcer (.098), D - White (.123), E - Munson (.119), F - Woods
     (.084), G - Ellis (.079), H - Michael (.064), I - Pitcher (.044),
     (Stottlemyre, Peterson, Kekich, Kline).

Team B. New York Yankees (All-time): AA - Rizzuto (.105), BB - Dimaggio
     (.155), CC - Gehrig (.155), DD - Ruth (.172), EE - Mantle (.159),
     FF - Lazzeri (.127), GG - Dickey (.120), HH - Rolfe (.102), II -
     Pitcher (.041), (Ford, Chesbro, Ruffing, Gomez).

Table II

The Outcome of the 10,000-Game Simulations

Run	Winning lineup	n_1	Losing lineup	$N-n_1$	$n_1-\mu$	$\|n_1-\mu'\|$	Hypothesis*
1	ATR	5008	ATR	4992	8		Accepted
2	ADX	5049	ADX	4951	49		Accepted
3	ADXR	5000	ADXR	5000	0		Accepted
4	ATR	5065	ADX	4935	65		Accepted
5	ATR	5123	ADXR	4877	123		Rejected
6	ADX	5056	ADXR	4944	56		Accepted
7	BTR	5055	BTR	4945	55		Accepted
8	BDX	5027	BDX	4973	27		Accepted
9	BDXR	5033	BDXR	4967	33		Accepted
10	BTR	5032	BDX	4968	32		Accepted
11	BTR	5130	BDXR	4870	130		Rejected
12	BDX	5101	BDXR	4899	101		Rejected
13	BTR	8214	ATR	1786		0	Accepted
14	BTR	8262	ADX	1738		48	Accepted
15	BTR	8428	ADXR	1572		214	Rejected
16	BDX	8174	ATR	1826		40	Accepted
17	BDXR	8066	ATR	1934		148	Rejected
18	HDD4	5057	HDD1	4943	57		Accepted
19	HDD4	5190	HDD9	4810	190		Rejected
20	HDD1	5098	HDD9	4902	98		Rejected
21	EDD4	5021	EDD1	4979	21		Accepted
22	HB4	5003	HB1	4997	3		Accepted

*that the outcome of baseball games is independent of batting order.

Summary

A Monte Carlo simulation of over 200,000 baseball games, using a programmed embodiment of the main features of the Sports Illustrated baseball
game, shows that batting order exerts only a small influence on the outcomes of baseball games. The effect of using the best batting order rather
than the worst is less than three extra wins per 162-game season. The
traditional lineup, wherein a team's strongest batters hit in the third
through fifth positions, is superior to a lineup in which batters are
arranged in decreasing order of productivity.

ANALYSIS OF BASEBALL AS A MARKOV PROCESS

by

Richard E. Trueman

Baseball can reasonably be viewed as a Markov process, since it meets the four basic requirements:

1. There are a finite number of possible outcomes, or states (here defined by the location of baserunners, the number of outs, and the lineup position).

2. The probabilities of moving from one state to another state, called transition probabilities, do not change over time; this is a slight simplification, since it ignores managerial strategies such as moving the infield or bringing in a pinch hitter. (Each transition probability is the probability of the given batter generating a particular play.)

3. The probability that the system will be in a given state depends only on the previous state and not how that state was reached; for example, a batter facing the situation of bases empty and no outs has the same probabilities, whether he is the first batter in the inning or follows a home run.

4. For each possible state, the probability that the system initially occupies that state is known; here, the only possible initial state is bases empty, no outs, and the leadoff batter up.

Baseball was discussed as an example of a mathematical model involving dynamic programming and Markov processes by Howard [1960] and Bellman [1964]. Both authors were primarily concerned with using baseball as an interesting example illustrating the usage of particular types of mathematical models. Howard, formulating a simplified computational example, went so far as to specify numerical values for the probabilities of a limited set of plays, but he assumed all players in the lineup were identical, and he issued a specific disclaimer regarding the validity of either assumptions or data. Bellman confined himself to the theoretical aspects of his models. Brown [1971] utilized Markov processes in an attempt to develop individual performance indices for various aspects of baseball.

In this paper, the intent is to show how a detailed model of baseball, considered as a Markov process, can be developed. Then, given actual baseball batting statistics for individual members of a specific team, different lineup orders are evaluated, and strategies analyzed for many different play situations. An alternative approach, using simulation, is described in a companion paper [Trueman, 1976].

The Mathematical Model. For each batter, visualize a 24 x 25 matrix which contains the transition probability from each of the 24 possible states (8 possible base conditions combined with 3 possible out conditions) to any other state, including the so-called trapping state representing 3 outs, the end of the half inning. It is assumed that, when a batter faces state i, the play results in a transition to state j, and the next batter is up. Thus, such plays as the stolen base, balk, wild pitch, runner picked off, etc., are omitted. There is no conceptual difficulty in ex-

tending the model to include such plays, but, in addition to lack of availability of much of the data needed, the size of the transition matrix would
be considerably enlarged; the state description would have to include not
only the location of baserunners but also their actual identification in
the given lineup, and so the number of states would be increased from 24
to several times that size.

The required transition probabilities for a given lineup are derived
from published individual batting statistics plus judicious estimates for
data required but not ordinarily available, such as a breakdown of types
of outs, singles and doubles by their effect on the movement of baserunners.
The thirteen plays used are described in [Trueman, 1976]. For convenience
in specifying the state transitions and in formulating the set of equations
to be developed later, an abbreviated set of play designations has also
been added. The plays are as follows (comments for outs apply for less
than two outs):

Play designation	Comment
NoAdv (no advance)	An out where no runners can advance: strikeout, popup, short fly, etc.
GdOut (ground out)	All runners advance one base.
Walk	Includes hit batsman.
PossDP (possible double play)	Double play if there is a runner on first base and less than two outs. The lead runner is out if he can be forced; runners not out advance one base. If no runner on first or two outs, this play is treated as NoAdv.
LFly (long fly)	Runner on third base scores.
OfSgl (outfield single)	A runner on first base advances one base; other runners advance a maximum of two bases.
LSgl (long single)	All runners advance a maximum of two bases.
HR (home run)	
IfSgl (infield single)	All runners advance one base.
VLFly (very long fly)	Runners on second or third advance one base.
Dbl (double)	Runners advance a maximum of two bases.
LDbl (long double)	All runners score.
Tpl (triple)	

Note: Singles also include a factor for errors allowing the batter
       to reach first.

Starting with the two-outs and bases-loaded state, and working back
to the no-outs and bases-empty state, it is possible to develop a set of
recursive equations to determine the probability that any given number of
runs will score during the remainder of the half-inning. By recursive,
we mean that each successive equation depends upon the results of one or
more equations developed earlier. We use the following simplified notation.

Let $R(n|k,i)$ = probability that n runs score in the remainder of the half-inning with the kth batter up facing the ith state; k takes on values from 1 to 9 and i from 1 to 24.

$P(d)$ = probability that the kth batter generates play d.

$m = k+1$ for $k \neq 9$; $m = 1$ for $k = 9$.

For clarity, each of the twenty-four values of i is denoted by four numbers. The first three represent the occupancy of first, second, and third base, respectively, and the fourth equals the number of outs. Thus, the combination 101:1 represents runners on first and third with one out. The trapping state is xxx:3, with the x's indicating that the location of baserunners is immaterial. The value of d will just be an abbreviated play designation previously specified or a readily understandable combination of plays.

The initial equation gives the probability that no runs will score with batter k facing the situation of bases loaded with two outs, or state 111:2.

$R(0|k,111:2) = P(\text{AnyOut})$. "AnyOut" means NoAdv + GdOut + PossDP + LFly + VLFly

No runs will score if the batter makes any type of out.

The next three equations are for runners on second and third, first and third, and first and second, respectively, all with two outs. In each case, the desired probability depends on $R(0|k,111:2)$ which has already been computed. For example:

$R(0|k,011:2) = P(\text{AnyOut}) + P(\text{Walk}) \times R(0|m,111:2)$

No runs score if the batter makes an out or if he draws a walk and there is no score when the next batter faces a bases-loaded situation. Here is where the recursive nature of the equations enters in. The calculations here for runners on second and third are dependent on the earlier calculations (which would have been done for all values of k) for the bases loaded.

$R(0|k,101:2) = P(\text{AnyOut}) + P(\text{Walk}) \times R(0|m,111:2) = R(0|k,011:2)$.

As can be seen, this probability would be identical to that for the situation with runners on second and third, as shown.

$R(0|k,110:2) = P(\text{AnyOut}) + [P(\text{Walk}) + P(\text{IfSgl})] \times R(0|m,111:2)$

Here, the next batter faces a bases-loaded situation if this batter gets a walk or hits an infield single. Any other non-out play would score at least one run.

The remaining four equations for the two-outs situation involve a runner on third, a runner on second, a runner on first, and bases empty. For example:

$R(0|k,001:2) = P(\text{AnyOut}) + P(\text{Walk}) \times R(0|m,101:2)$

Continuing with the two-outs situations, the probability that exactly one run scores in the remainder of the half inning, when the bases are loaded, is:

$$R(1|k,111:2) = P(Walk + IfSgl) \times R(0|m,111:2)$$

> If just one run is to score, the batter must either walk or hit an infield single to score one run, and no runs can then score when the next batter faces a bases-loaded situation.

At this point, the general approach should be reasonably clear, so let us now discuss how these recursive equations, once developed, can be utilized.

Uses of the Model. The complexity of this model and the thousands of calculations required obviously mandate the development of a computer program. Such a program can perform extensive analyses of a given lineup in a few seconds of computer time.

Basically, the computer program operates as follows:

1. Read in the individual batting statistics for the given team.

2. For the chosen lineup and batting order, convert the batting statistics to an ordered set of play probabilities, as provided for in the model.

3. For every batter, calculate, for each of the 24 base-and-out situations, the probability of scoring no runs in the remainder of the half-inning.

4. Evaluate strategies used attempting to score at least one run, by computing the required breakeven success probability [Trueman, 1976].

5. For every batter, calculate, for each of the 24 base-and-out situations:

    a. the probability of scoring 1 run in the remainder of the half-inning. Repeat these calculations for 2, 3,...,14 runs (14 is an arbitrary upper limit).

    b. the expected number of runs scored in the remainder of the half-inning.

6. If desired, evaluate strategies used attempting to increase the expected number of runs scored, by calculating the required breakeven success probabilities.

7. Calculate the probability distribution of the number of batters in an inning, and then calculate, for every inning, the probability that each batter leads off.

8. Compute the expected number of times each batter leads off an inning in a nine-inning game and multiply by the expected number of runs scoring when he leads off (faces a bases-empty, no-outs situation). Accumulate these values to get the total expected runs per game for this lineup.

To illustrate the use of the analytical model described, and to permit comparisons with the simulation model, we use the same lineup of the 1973 Los Angeles Dodgers as [Trueman, 1976]. The batting statistics for this lineup are given in Table I.

Table I

A 1973 Los Angeles Dodgers Lineup

	AB	H	2B	3B	HR	RBI	SH	SF	SB	CS	W	HP	SO	GIDP	PCT
Lopes	535	147	13	5	6	37	7	6	36	16	56	5	77	14	.275
Mota	293	92	11	2	0	23	6	2	1	3	16	1	12	11	.314
Davis	599	171	29	9	16	77	1	7	17	5	17	5	62	7	.285
Ferguson	487	128	26	0	25	88	0	10	1	1	78	1	81	5	.263
Garvey	349	106	17	3	8	50	0	3	0	2	7	3	42	8	.304
Cey	507	124	18	4	15	80	4	8	1	1	67	2	77	19	.245
Crawford	457	135	26	2	14	66	3	4	12	5	66	1	91	6	.295
Russell	615	163	26	3	4	56	9	7	15	7	14	1	63	9	.265
Pitcher	423	64	8	0	0	24	40	1	0	0	21	0	100	12	.151

Analysis of Strategies. In order to analyze strategies where the goal is to increase the probability of scoring at least one run, we need to know, for every batter, the probability of one or more runs scoring in the remainder of the half inning, for specific base-and-out situations which he may face. This information is readily obtained from the data of Table II, which shows, by batter and situation, the probability of not scoring in the remainder of the half inning. The breakeven success probabilities for different strategies are then calculated, and are shown in Table III.

The strategies analyzed in Table III are two of those most frequently used, the sacrifice and the stolen base. (The squeeze play is just a sacrifice with a runner on third.) Based on the data in this table, the following general observations (which are believed to be applicable to any representative major-league baseball lineup) can be made:

1. The attempted sacrifice, in general, is a very poor strategy.

    a. In the usual sacrifice situation, with a man on first and no outs, only the pitcher should be asked to bunt, although one or two others in the lineup, if they are excellent bunters, could reasonably attempt to sacrifice upon occasion.

    b. If a good bunter, the pitcher could attempt the sacrifice with a runner on first and one out.

    c. With one out, the sacrifice is never worthwhile if there is a man on second or men on first and second.

    d. For batters who can sacrifice successfully three-fourths of the time, the squeeze play with a runner on third and one out should be considered.

    e. Even in those cases where the probability of scoring at least one run is increased, the sacrifice always reduces the expected number of runs.

Table II
Probability No Runs Score, by Situation

Two-outs Situations

Batter	111	011	101	110	001	010	100	000
Lopes	.634	.698	.698	.727	.702	.736	.868	.941
Mota	.631	.667	.667	.704	.668	.708	.876	.944
Davis	.673	.694	.694	.720	.698	.728	.864	.920
Ferguson	.621	.711	.711	.731	.714	.739	.840	.906
Garvey	.660	.678	.678	.708	.679	.714	.870	.928
Cey	.650	.720	.720	.740	.731	.757	.863	.922
Crawford	.597	.686	.686	.714	.689	.721	.856	.936
Russell	.701	.719	.719	.754	.720	.757	.923	.970
Pitcher	.789	.819	.819	.837	.822	.844	.934	.970

One-out Situations

Batter	111	011	101	110	001	010	100	000
Lopes	.388	.354	.424	.546	.356	.511	.725	.834
Mota	.259	.240	.275	.504	.243	.495	.683	.815
Davis	.209	.230	.242	.515	.232	.519	.656	.797
Ferguson	.225	.258	.270	.492	.262	.519	.665	.805
Garvey	.286	.286	.315	.537	.292	.528	.706	.822
Cey	.334	.341	.392	.548	.347	.536	.706	.833
Crawford	.270	.296	.313	.509	.299	.538	.719	.868
Russell	.295	.300	.313	.624	.302	.623	.794	.902
Pitcher	.429	.434	.477	.638	.438	.614	.788	.891

No-outs Situations

Batter	111	011	101	110	001	010	100	000
Lopes	.073	.123	.082	.350	.125	.336	.561	.685
Mota	.045	.079	.056	.302	.081	.296	.495	.672
Davis	.061	.083	.076	.286	.085	.298	.482	.680
Ferguson	.082	.105	.103	.299	.109	.326	.504	.681
Garvey	.093	.133	.111	.351	.136	.335	.536	.719
Cey	.077	.144	.100	.350	.147	.368	.573	.743
Crawford	.088	.123	.109	.346	.126	.380	.581	.760
Russell	.120	.158	.137	.427	.160	.410	.620	.797
Pitcher	.176	.214	.196	.432	.216	.401	.629	.772

2. With good (not necessarily outstanding) base stealers, the attempted stolen base can be worthwhile in several different situations.

   a. The attempted steal with a man on first is most favorable with two outs, while the one-out situation appears to be slightly preferable to that with no outs.

Table III

Analysis of Strategies Used Attempting to Score at Least One Run

Attempted Strategy	Breakeven Success Probability, Percent								
Batter	1	2	3	4	5	6	7	8	9
Sacrifice - man on 1st, no outs	65	*	*	*	*	81	*	97	45
Sacrifice - man on 1st, 1 out	90	*	*	*	*	*	*	*	60
Sacrifice - man on 2nd, no outs	63	78	86	85	*	71	76	*	71
Sacrifice - man on 2nd, 1 out	*	*	*	*	*	*	*	*	*
Squeeze play - man on 3rd, no outs	82	88	87	85	81	80	84	80	70
Squeeze play - man on 3rd, 1 out	59	72	72	70	66	60	68	68	50
Sacrifice - men on 1st and 2nd, no outs	58	75	88	95	95	75	86	*	59
Sacrifice - men on 1st and 2nd, 1 out	*	*	*	*	*	*	*	*	*
Steal - man on 1st, no outs	55	62	63	63	59	56	59	57	54
Steal - man on 1st, 1 out	50	58	66	62	56	56	54	51	51
Steal - man on 1st, 2 outs	50	42	50	61	46	56	52	32	42
Steal - man on 2nd, no outs	70	71	70	69	71	68	66	66	73
Steal - man on 2nd, 1 out	74	64	58	60	63	67	62	52	67
Steal - man on 2nd, 2 outs	89	88	90	91	89	90	90	87	88
Steal - man on 3rd, no outs	85	90	89	86	83	82	85	82	76
Steal - man on 3rd, 1 out	62	74	75	71	69	62	68	69	55
Steal - man on 3rd, 2 outs	30	33	30	29	32	27	31	28	18
Double steal - men on 1st and 2nd, no outs	41	46	53	53	45	47	46	42	45
Double steal - men on 1st and 2nd, 1 out	50	44	43	51	41	50	50	29	50
Double steal - men on 1st and 2nd, 2 outs	91	89	92	93	91	93	91	87	90

Note: * indicates a strategy which can never be favorable even if always 'successful.

---

b. With a man on second, the attempted steal is less favorable than with a man on first. For the two-outs situation with a man on second, the required success probabilities are prohibitively high.

c. With a runner on third and two outs, the attempted steal of home should be considered if it succeeds at least one-third of the time. This is by far the lowest required success probability for any situation considered.

d. Except for the two-outs case, the attempted double steal with men on first and second has a lower required success probability than for a man on first and the same number of outs. With a good runner on second base, this strategy should perhaps be attempted more frequently.

Analysis of lineup orders. For a given lineup order, the expected number of runs scored in nine innings can be determined as shown in Table IV.
For each batter, we calculate the expected number of times he leads off in
a nine-inning game (using another set of recursive equations not described
here) and multiply this value by the expected number of runs which score when
he faces the bases-empty, no-outs situation. These products, shown in the
final column of Table IV, represent the run contribution for each such 000:0
situation. Their total is the expected number of team runs in nine innings.
These run contribution values are associated with a given leadoff situation;
they do not represent the contribution of the specified batter.

The calculated value of 4.72 runs per nine-inning game is undoubtedly
high for this lineup, since the 1973 Dodgers averaged only about 4.12 runs
per nine innings. Some of this difference can be attributed to the fact
that the sacrifice is not included. From calculations performed by Cook
[1971], the sacrifice, as customarily employed, results in a decrease of
approximately 0.2 runs per game. In addition, this particular lineup had an
estimated batting average of .268, compared to the overall team average of
.263. Some or all of the remaining difference in run production can be attributed to the fact that the mathematical model described does not take into
account many of the plays which can and do happen in major league baseball,
particularly outs which occur on the bases (other than ground-into double
plays).

Since the primary interest here is in making relative comparisons between lineup orders, the model is still quite useful. For example, when
several different orders were tried for the given lineup, the most favorable
result was for one particular lineup which gave an expected value of 4.75
runs. The small difference between this value and that of 4.72 runs for the
lineup analyzed in detail (and actually used by the Dodgers) is not believed
to be significant, considering, among other things, the assumptions that had
to be made for certain play probabilities. In general, given a particular
lineup of nine players, managers appear to be doing quite a good job of selecting the most productive batting order. It is not likely that significant improvement in scoring potential can be achieved by changing the present approach of placing those hitters who get on base relatively frequently
at the top of the batting order, weak hitters at the bottom, and power hitters in the middle. This was corroborated by an extensive simulation analysis performed by Freeze [1974].

For a given lineup of nine players, and assuming that the pitcher will
always bat ninth, there are 8! or 40,320 possible orderings of the first
eight players. (With a designated hitter in the lineup, there are 9! or
362,880 possible orderings of the nine players.) Even with a high-speed computer, this represents an unrealistically large number of combinations to
evaluate. If this computer model were to be implemented for a major-league
team wishing to evaluate different lineup orders, heuristic rules could
surely be developed, based on some exploratory analyses, to test perhaps a
hundred of the most potentially promising lineup orders.

One use for a comparative analysis of different lineups would be for
the evaluation, from the standpoint of team run production, of potential
player trades. This would be especially valuable when considering a power
hitter vs. a high-average hitter with limited power. Also, if two players
at the same position differed considerably in their hitting and fielding
ability, the one being the better hitter and the other the better fielder,

Table IV

Calculation of Expected Team Runs

Batter	Expected times leads off	Expected runs scored facing 000:0 situation	Expected run contribution from this 000:0 situation
Lopes	2.006	.624	1.25
Mota	.799	.620	.50
Davis	.727	.581	.42
Ferguson	1.036	.555	.57
Garvey	.930	.482	.45
Cey	.921	.454	.42
Crawford	.913	.433	.40
Russell	.778	.391	.30
Pitcher	.889	.461	.41

Expected team runs for 9 innings   4.72

the offensive contribution of the better hitter could be quantitatively measured and weighed against his defensive shortcomings, if they could be quantified in terms of the expected increase in opponents' runs scored.

Summary

By considering the offensive aspects of baseball as a Markov process, it becomes possible to develop a comprehensive mathematical model which is computationally efficient. This model can be effectively utilized in the analysis of many different types of strategies and in the comparative evaluation of different lineups and batting orders. When and if actual baseball statistics become available on some of the more detailed aspects of baseball play, the model can readily be expanded to any desired level of complexity. Detailed analyses of an actual major-league lineup, especially those involving the required breakeven success probabilities of a number of strategies, demonstrate the potential value of such models to baseball managers and top management.

DYNAMIC PROGRAMMING AND MARKOVIAN DECISION
PROCESSES, WITH APPLICATION TO BASEBALL

by

Richard Bellman

Based on an article in "Applied Combinatorial Mathematics", 1964.

Introduction. Expansion of big-league baseball in 1977 has revived, at least in the minds of aficionados, a long-standing question: "Are managers really necessary?" Do they exist solely to enforce curfew and to provide copy for sportswriters, or do they possess a crystal ball and a dowsing rod that enable them to make the right decisions at the right time?

Since baseball is a multistage decision process of stochastic type, as will be explained later, we can go a long way toward answering this question with the help of the theory of dynamic programming and digital computers.

We shall describe various formulations in some detail and then examine some of the analytic and computational aspects, none of which are trivial. Numerical examples of these principles are found in articles in this book by Howard, Trueman, and others.

Baseball as a Multistage Decision Process; State Variables. To simplify the discussion, let us begin with an analysis of the problems confronting the manager of the team at bat. Later, we shall consider the full problem, in which attention must be paid to the characteristics of the team in the field, to the particular properties of the pitcher, to the arrangement of the fielders, etc. Initially, let us average over different pitchers, different defensive arrangements, etc., and assume therefore that we are facing an average team--for example, an average pitcher, who delivers a strike with a certain probability p and a ball with a certain probability 1 - p.

With this as the background, what data does the manager require to carry out his tactical and strategic maneuvering? The full information pattern is the following·

1. The score.

2. The inning.

3. The batter.

4. The number of outs.

5. The count on the batter, that is, the number of balls and strikes.

6. The men on base and their location.

Let us first ignore the inning, which is to say we are initially contemplating the early stages of the game, and also let us ignore the score. We shall discuss these assumptions in more detail below.

We now proceed to enumerate the possible situations that can arise in

77

a particular inning as far as the team at bat is concerned.

There are three possibilities for the number of outs, four possibilities for the number of balls, three possibilities for the number of strikes, eight possible ways in which there can be 0, 1, 2, or 3 men on base, nine possible batters (supposing, as we shall, that pinch hitters will not be used in the early innings). There are then

$$3 \times 4 \times 3 \times 8 \times 9 + 1 = 2593$$

different possible situations that can occur within an inning, where the extra one is for the three-out situation. Were we to take into account the inning, an additional factor of nine, this would increase the number to 23,337, whereas variations in scores would increase this total to a number in excess of 10^5, a respectably large number.

Let us for the moment denote these possibilities, in no particular order, by the symbol i, i = 1,2,...,2593, and call them the states of the system. Subsequently, we shall point out that for conceptual and computational purposes, some arrangements are far preferable to others. The preferential labelings arise very naturally from the structure of the game, as is to be expected.

Decisions. We shall assume that the manager is responsible for every action taken by the batter and the base runners. In each of the foregoing situations, we suppose that he signals the players, and that they then follow his instructions as best they can.

In each state, the fundamental initiating action is the delivery of a pitch by the pitcher. The batter has the prerogative of taking the pitch without swinging at it, of swinging and attempting to hit the ball, or of attempting a bunt. The men on base have the prerogative of attempting to steal, or not.

The manager then has the responsibility of deciding among the following alternatives·

    1. The batter "takes" the next pitch.

    2. The batter swings at the next pitch if it is a strike.

    3. The batter swings at the next pitch in any event, that is, he carries out a hit-and-run play.

    4. The batter attempts a bunt.

    5. Men on base are instructed to steal.

As a result of these decisions, certain events can take place·

    1. The batter has a ball or strike added to his count, which for three balls or two strikes may mean respectively a walk or a strikeout.

    2. The batter attempts to hit the ball and misses--a strike and possibly an out.

    3. The batter hits a foul--a strike or not, depending on the count.

4.   The batter swings and hits a fair ball; this may
     be a single, double, or triple out, and it may be
     a hit--a single, double, triple, or homer.

Let us enumerate the possible decisions D in some fashion and denote
the set of all decisions by {D}.

In each of these eventualities, a certain number of runs may be scored,
depending on the type of hit or out and the location of the runners. Observe, however, that whatever happens, we return to one of the 2593 possible
situations enumerated in the previous section.

A characteristic feature of baseball is that it is a stochastic process. Starting in a given state and having made a particular decision, we
cannot possibly predict exactly what the next state is going to be. We
can, however, determine the set of possible states that can result from a
given decision in a particular state.

Criterion Function. Let us now examine various ways of evaluating the
decisions that are made. Our over-all objective, of course, is to win the
game. Bearing in mind that this is a stochastic event, the two most immediate measures are the probability of scoring at least one run and the probability of maximizing the expected number of runs scored. Clearly these
are not equivalent criteria.

Both objectives possess important invariance properties that greatly
aid the mathematical determination of optimal strategies. With either criterion, regardless of what has transpired in the past, one continues from
the present state according to the same criterion. Strategy based on the
probability of winning requires the additional state variables of the score
and the inning, whereas strategies based on the two immediate measures mentioned above do not require this information, an advantage that we shall exploit to reduce the number of state variables to a manageable size.

It is a reasonable approximation to suppose that in the early innings
a team will play so as to maximize the number of runs that it scores in an
inning, whereas in later innings it will play to either maximize the probability of scoring a run (protecting a lead) or maximize the probability
of at least tying the score.

Policies, Optimal Policies, and Combinatorics. In any particular situation, specified by a state i, i = 1,2,...,2592, we have a choice of any
of a number of possible decisions. Since the decision D that is made is
clearly dependent on the state i, it is a function of i, which we shall denote by D(i). Any function of this type, which maps the set of states onto
the set of decisions, is called a policy. A policy that maximizes the criterion function is called an optimal policy.

The determination of winning strategies is equivalent to the determination of optimal policies.

The problem of obtaining optimal policies can in turn be conceived of
as a combinatorial question along the following lines. Each sequence of
events in an inning can be represented by a set of integers $(i_1, i_2, ..., i_k)$.
Let us arrange the states in a linear order. Then the sequence of events
can be represented by a generalized random-walk process in which the transi-

tion probabilities from state to state change over time in accordance with
the decisions that are made.

A priori, since each decision has a number of possible consequences,
we have a stochastic graph. It is theoretically possible to obtain policies by tracing out the consequences of various policies and averaging over
the outcomes of these alternatives. This would be a laborious process, in-
elegant at best, and impossible to execute computationally. The number of
possible paths increases in an alarming way.

Generally speaking, we cannot handle combinatorial processes on a computer in a routine fashion. To encompass the enormous numbers of cases
that arise in dealing with even the simplest problems, we must consider the
structure of the problem; that is, we must introduce sophisticated mathematical techniques.

One systematic technique is to transform combinatorial problems into
analytic problems that are amenable to the classical methods of analysis.
Can we perform this metamorphosis here in order to obtain optimal policies
without actual enumeration?

Principle of Optimality. To accomplish this transformation, we employ
the theory of dynamic programming, relying on the following fundamental
characterization of optimal policies: An optimal policy has the property
that whatever the initial state and initial decisions are, the remaining decisions must constitute an optimal policy with regard to the state resulting from the decisions already made. This simple and intuitive statement
leads to a functional equation that permits us to determine optimal policies by analytic means.

Functional Equations. Let us begin with the obvious remark that the
expected number of runs scored starting in a particular state i depends on
this state. Define therefore the function

$$f(i) = \text{expected number of runs scored in an inning,} \qquad (1)$$
$$\text{starting in state i and employing an optimal}$$
$$\text{policy.}$$

A decision D in state i results directly in an expected number of runs
scored $r_i(D)$ and, with probability $p_{ij}(D)$, a transition into state j. The
principle of optimality, cited in the foregoing section, asserts that if
decision D is made in state i, and thereafter an optimal policy is followed,
then the expected number of runs to be scored in the rest of the inning is
given by the expression

$$r_i(D) + \sum_j p_{ij}(D) f(j) \qquad (2)$$

Since this holds for any choice of D, clearly an optimal policy D is chosen
so as to maximize (2). Hence, we have

$$f(i) = \max_D \left[r_i(D) + \sum_j p_{ij}(D) f(j) \right] \qquad (3)$$

an equation determining optimal baseball strategy. Although we hardly expect Casey Stengel to solve equations of this nature, it may be that we can
obtain significant results with the aid of digital computers.

Existence and Uniqueness. Before we devote time to the analytic and
computational solution of equations of this type, it is essential that we
make sure that the equation has a solution, and that either it has a unique
solution or we know how to focus on the solution that corresponds to the
multistage decision process we are examining.

The general study of existence and uniqueness of equations of this
nature leads to some very difficult problems. In this case, however, using
the analytic device discussed in the following section, we readily obtain
the desired results. They are simple consequences of the fact that any
policy has nonzero probability of leading to the "trap" state--three outs
in this case.

Analytic Aspects. Consider now the question of obtaining an exact
analytic expression for f(i), and thus a direct determination of the optimal policy. To illustrate the basic idea, let us start with the one-dimen-
sional version, the scalar equation

$$u = \max_q \left[a(q) + b(q)u \right] \qquad (4)$$

where to avoid extraneous questions of continuity we suppose that q ranges
over a discrete set S of values, that the function a(q) is finite on this
set, and that $0 \le b(q) < 1$ on this set. Provided that (4) has a solution,
we have the relation

$$u \ge a(q) + b(q)u \qquad (5)$$

for each $q \in S$, and therefore

$$u - b(q)u \ge a(q) \qquad (6)$$

or

$$u \ge \frac{a(q)}{1 - b(q)} \qquad (7)$$

Since equality holds for one q, we have

$$u = \max_q \left[\frac{a(q)}{1 - b(q)} \right] \qquad (8)$$

On the other hand, the method of successive approximations, or a fixed-
point method, establishes the existence of a solution. As the reader can
verify, one can go directly from (8) to (4).

Can we extend this explicit representation to the vector-matrix case?
Suppose that

$$X = \max_q \left[A(q) + B(q)X \right] \qquad (9)$$

where q again ranges over a discrete set, X is an n-dimensional vector, as

is A(q), and B(q) an n x n matrix with nonnegative elements. Then, as previously, we have

$$X \geq A(q) + B(q)X \tag{10}$$

for all q ϵ S, and

$$[I - B(q)]X \geq A(q) \tag{11}$$

where I is the n x n identity matrix. The problem thus reduces to an examination of the conditions under which (11) implies

$$X \geq [I - B(q)]^{-1}A(q) \tag{12}$$

This question is part of the general theory of positive operators. Matrices
possessing this positivity property arises naturally in probability theory
and mathematical economics and have been the subject of much research.

It is clear that (12) is a consequence of the nonnegativity of the
elements of the matrix $[I - B(q)]^{-1}$. A simple and meaningful condition that
guarantees this is

$$\sum_j b_{ij}(q) < 1, \quad i = 1, 2, \ldots, n, \tag{13}$$

with $b_{ij}(q) \geq 0$ for all i and j. Condition (13) is not directly satisfied
in our case, but there is no difficulty in circumventing this obstacle.

Under the foregoing hypothesis, we can write the solution of (9) in
the form

$$X = \max_q [I - B(q)]^{-1} A(q) \tag{14}$$

In this case, approximate methods bypassing matrix inverses are more useful and, as we shall see, more significant relative to the actual process.

Direct Computational Approach. Let us now examine a direct computational approach to the solution of (3) based on the classical method of
successive approximations. Write

$$f_0(i) = \max_D r_i(D)$$

$$f_{n+1}(i) = \max_D \left(r_i(D) + \sum_j p_{ij}(D)f_n(j) \right), \quad n = 0,1,2,\ldots \tag{15}$$

The interpretation of these equations, as far as the original process is
concerned, is quite direct. The first function $f_0(i)$ represents the maximum average number of runs scored on a particular play, whereas $f_n(i)$ represents the maximum expected number of runs scored under the constraint
that at most $n+1$ additional men will come to bat. Since the probability
of going through the batting order in a particular inning is quite small,
we see that $f_8(i)$ should be an excellent approximation to $f(i)$, and probably even $f_5(i)$ will also be in good agreement.

The foregoing approach avoids discussions of existence and uniqueness of solution. If, however, one is interested in obtaining $f(i)$ rather than $f_8(i)$, it is worth noting that the convergence is monotone and geometric; that is, we have

$$f(i) - f_n(i) \leq c_1 r^n, \quad 0 < r < 1, \quad c_1 > 0 \tag{16}$$

Now let us examine the storage aspects. To compute $f_{n+1}(i)$, we must store $f_n(i)$ and the quantities $r_i(D)$ and $p_{ij}(D)$. Since $1 \leq i \leq 2592$, we see that each matrix $[p_{ij}(D)]$ involves $2592 \times 2593 = 6,721,056$ entries, which is a sizable number insofar as rapid access storage is concerned. What saves the situation is the fact that most of the $p_{ij}(D)$ are zero, regardless of the choice of D. In other words, from any particular state i one can go to only a small number of other states.

Stratification. In many processes of this nature, some parts of the problem can be solved independently of others: natural decompositions enable us to determine the optimal policy in pieces. In the present case, one such decomposition is clear. First we can determine what to do when there are two out. Having calculated those values $f(i)$ involving two outs in the state variable, we can then calculate the values of $f(i)$ involving one out, using the previously computed values. Once this is done, we turn to the calculation of those values of $f(i)$ involving no outs. The computational advantage of doing this is great, since sometimes we can reduce an involved problem, for which the storage requirements when considered in a straightforward fashion exceed the capabilities of modern computers, to one that is quite tractable.

The theoretical importance of stratification is that we can often use it to transform a formidable problem, in which a condition of the type

$$\sum_j p_{ij}(D) < 1 \tag{17}$$

does not hold for all i and all D, into a sequence of subproblems within each of which a condition of this nature--which, as pointed out, ensures nonnegativity of $[I - P(D)]^{-1}$--does hold.

This is the case here, since with two out all "swing" decisions involve a nonzero probability of an out and thus an end to the inning. Some of the "take" decisions, however, do not immediately involve the probability of an out. By means of successive elimination of these states, and use of (5) to (8), we can reduce all remaining equations to those for which (17) holds.

Approximation in Policy Space. In what has preceded, we have employed the classical method of successive approximations with some small modifications. Let us now turn to a method of successive approximations that has no counterpart in classical analysis, the technique of approximation in policy space. Returning to (2), let us begin by guessing an initial policy $D_0(i)$ and compute the return function $f_0(i)$ obtained on the basis of this policy. The return function $f_0(i)$ satisfies the functional equation

$$f_0(i) = r_i(D_0) + \sum_j p_{ij}(D_0) f_0(j) \tag{18}$$

an equation that we can solve by direct iteration--or perhaps, if we wish
to avoid some storage difficulties, by Monte Carlo methods. We now determine a superior policy $D_1(i)$ by carrying out the maximization

$$\max_{D} \left[r_i(D) + \sum_j p_{ij}(D)f_o(j) \right] \quad , \quad \text{for } i = 1,2,\dots \tag{19}$$

Having obtained $D_1(i)$ in this fashion, we compute the new return function $f_1(i)$ by means of the equation

$$f_1(i) = r_i(D_1) + \sum_j p_{ij}(D_1)f_1(j) \tag{20}$$

It is not difficult to establish rigorously the inequalities

$$f_o(i) \leq f_1(i) \leq \dots$$

for all i. Approximation in policy space yields monotone convergence, a
most important property.

Utilization of Experience and Intuition. A most important attribute
of approximation in policy space is that it allows us to take maximum advantage of actual experience in carrying out the process. Consider, for
example, the game of baseball. In many of the 2592 situations that might
arise, the optimal decision is obvious. Consequently, we start immediately
with the correct value of $D(i)$, the optimal policy, for many values of i.
This excellent first guess speeds up the calculation of the solution to a
vast extent. Furthermore, it should be emphasized that human beings think
automatically in terms of policies, and not immediately in terms of return
functions.

Two-Team Version of Baseball. The actual game of baseball involves
two teams that are constantly maneuvering and countermaneuvering to gain
an advantage. This means that the determination of optimal baseball
strategy requires the analysis and computational solution of multistage
games. There is no difficulty in applying dynamic programming to the analytic formulation of processes of this nature. The computing time per
stage goes up a little but not drastically. Since there is no difficulty
in principle, we shall bypass this excursion, however inviting, and turn
to some other matters.

Probability of Tying or Winning. Suppose that we are interested in
determining optimal play in connection with tying the score. We then introduce the functions

$$f_k(i) = \text{probability of scoring at least k runs, starting} \tag{22}$$
$$\text{in state i and using an optimal policy, } k = 1,2,\dots$$

The outcome of a decision D will, with probability $p_{ijr}(D)$, be a new
state j and a certain number r of runs scored. Hence, the equation for
$f_k(i)$, obtained from the principle of optimality, is

$$f_k(i) = \max_{0 \le r \le k} \sum \sum_j p_{ijr}(D) f_{k-r}(j)$$

with $f_0(i) = 1$. These equations are quite easy to solve computationally, starting with those for $f_1(i)$, as an application of the stratification technique mentioned above.

Equations of this nature will be particularly useful in analyzing strategy in late-inning and extra-inning play.

Discussion. In the foregoing, we have assumed that the process is steady-state, by which we mean that the process stays the same not only from day to day but also from hour to hour. Fortunately, this is not true. Athletes do not perform as well on different days or at different times on the same day. It takes the experienced eye of a manager to tell if an athlete is in good condition, mentally as well as physically.

As in poker, one can do very well playing percentage; but, if one wants to win big, one has to play psychology.

COMPARING THE RUN-SCORING ABILITIES OF
TWO DIFFERENT BATTING ORDERS: RESULTS OF A SIMULATION

by

Arthur V. Peterson, Jr.

Which is a more productive batting order for a baseball team: one
in which the two best hitters bat in the third and fourth positions in
the order (a standard batting order), or one in which the two best hitters
bat first and second (a best-first batting order)? Each appears to have
its advantages. For example, in a standard batting order the best hitters
have more good opportunities to bat in runs, because the batters who pre-
cede them in the batting order are good hitters and therefore tend to get
on base often. On the other hand, in the best-first batting order, the
best hitters have more appearances at bat. When all the factors are considered, which batting order has the greater run-scoring ability?

Results that shed some light on this question were obtained by simu-
lating one hundred 162-game seasons twice--once with a standard batting
order, and then again with a best-first batting order. The probabilities
of the various events (outs, singles, doubles, etc.) for the nine hitters
used in the simulation were chosen based on actual frequencies of these
events for typical major league hitters. The results, all of which are
averages per season of data gathered for 100 simulated seasons for a
typical major-league team (not including designated hitter), are briefly
summarized in Table I.

For the standard batting order:

1. Over 30 per cent (197) of the team's runs were scored by
    batters C and D, the third and fourth batters.

2. Almost as many (187) were scored by A and B, the first
    and second batters.

3. Over one-third (208) of the runs batted in were attributed
    to batters C and D.

4. The two weakest batters, H and P, together scored only 78
    runs, less than 15% of the total.

5. Overall, the team scored 639 runs.

For the best-first batting order:

1. Over 37% (241) of the team's runs were scored by batters
    C and D, now the first and second batters in the batting
    order. They scored 44 more runs than they scored in the
    standard batting order.

2. Only 22% (141) of the runs were scored by batters A and B,
    now in the third and fourth places in the batting order. They
    scored 46 fewer runs than they scored in the standard batting
    order.

This work was partially supported by NIH Grant No. 5TI-GM-25-17.

86

3. The runs batted in are rather evenly distributed among the first 4 batters, whereas in the standard batting order batters C and D batted in almost twice as many runs (208 vs. 110) as batters A and B.

4. Again, the least contribution was made by the two weakest batters, H and P. But in this batting order they managed to score 9 more runs, 87 vs. 78, than they scored in the standard batting order.

5. Overall, the team scored 646 runs, 7 more than it scored with the standard batting order

Table I

Batting order	Batting average	On-base average	Home runs	Appearances	Runs scored	Runs batted in
Standard batting order (100 seasons):						
A	.290	.359	11	764	99	52
B	.270	.333	7	746	88	58
C	.268	.412	44	728	112	108
D	.335	.404	20	712	85	100
E	.262	.330	9	698	61	79
F	.249	.324	14	681	59	78
G	.220	.305	22	663	57	77
H	.220	.296	3	644	42	43
P	.163	.207	0	625	36	24
Team Totals	.255	.333	130	6261	639	619
Best-first batting order (100 seasons):						
C	.268	.412	47	764	135	89
D	.334	.404	21	747	106	77
A	.290	.358	10	731	78	82
B	.273	.334	6	715	63	85
E	.262	.330	9	698	61	77
F	.249	.324	14	681	59	75
G	.220	.305	22	663	57	75
H	.220	.296	3	644	45	42
P	.163	.207	0	625	42	24
Team Totals	.255	.334	132	6268	646	626

These results appear to indicate that the best-first batting order would be more productive than the standard batting order for this particular team. But two cautions should be made against drawing firm conclusions from these results. First, these results do not pertain to teams other than the specific one for which the simulation was performed, and other teams [Cook, 1971; Freeze, 1974] yield slightly different

results. Secondly, even for our specific team, 100 seasons are too few
in this case to draw conclusions on the true difference in run-scoring
ability with high confidence. The judgment that the best-first batting
order is a more productive one than the standard batting order, based on
a difference of 7 runs compared to a standard error of 3.8 runs, is
necessarily a tentative one.

The difference in run scoring between the two batting orders (7 runs
per season) for this team was small compared with the total number of runs
scored per season (over 600 runs). It is the author's judgment that this
difference, and therefore any conclusion about the relative productiveness
of different batting orders, depends heavily on the specific players and
the specific team.

ACKNOWLEDGMENT

The author gratefully thanks Mark J. Cannelora, Thomas M. Cover, and
Marjorie F. Peterson for their helpful suggestions.

Summary

A simulation was performed to compare the run-scoring abilities of
two different batting orders for a typical major league team. The results
were that, for this team, a best-first order was slightly more productive
than a standard batting order. However, the advantage was small. The
author's judgment is that any conclusions about the relative productiveness of different batting orders depend heavily on the specific players
and the specific team.

BASEBALL A LA RUSSE

by

Ronald A. Howard

Based on a chapter in <u>Dynamic Programming and Markov Processes,</u> 1960

In making any decision, it is tempting, though hazardous, to focus on the immediate rather than the long-run effects. As an illustration, a novice baseball manager might be tempted to emphasize the runs to be expected from the next play at the expense of the base-and-out position resulting from the play. The Markov decision process model [Howard, 1960] allows us to investigate this problem in simplified form.

Consider an early half-inning of a baseball game. The teams are unusual because all players are identical in athletic ability and their play is unaffected by the tensions of the game. The manager makes all decisions regarding the strategy of the team, and his alternatives are limited in number. He may tell the batter to hit or bunt, tell a man on first to steal second, a man on second to steal third, or a man on third to steal home. For each situation during the inning and for each alternative, there will be a probability of reaching each other situation that could exist and an associated reward expressed in runs. Let us specify the probabilities of transition under each alternative as shown in Table I.

The state of the system depends upon the number of outs and upon the situation on the bases. We may designate the state of the system by a four-digit number $d_1 d_2 d_3 d_4$, where d_1 is the number of outs--0,1,2, or 3--and the digits $d_2 d_3 d_4$ are 1 or 0 corresponding to whether there is or is not a player on bases 3, 2, and 1, respectively. Thus the state designation 2110 would identify the situation "2 outs; players on second and third," where 1111 would mean "1 out; bases loaded." The states are also given a decimal number equal to $1 + 8d_1 +$ (decimal number corresponding to binary number $d_2 d_3 d_4$). The state 0000 would be state 1, and the state 3000 would be state 25; 2110 corresponds to 23, 1111 to 16. There are eight base situations possible for each of the three out situations 0, 1, 2. There is also the three-out case where the situation on base is irrelevant; we may arbitrarily call this the state 3000. Therefore, we have a 25-state problem.

The number of alternatives in each state is not the same. State 1000 or 9 has no men on base, so that none of the stealing alternatives are applicable, and only the hit or bunt options are present. State 0101 or 6 has four alternatives: hit, bunt, steal second, or steal home. State 3000 or 25 has only 1 alternative, and that alternative causes it to return to itself with probability 1 and reward 0. State 25 is a trapping or recurrent state; it is the only state that the system may occupy as the number of transitions becomes infinite.

The following notation is helpful. We let p_{ij}^k be the probability of a transition from state i to state j if alternative k in state i is used. The reward from the transition, in runs, is r_{ij}^k. The expected immediate reward from a transition from state i is then $q_i^k = \sum_{j=1}^{N} p_{ij}^k r_{ij}^k$.

Table I. Data

1. Manager tells player at bat to try for a hit.

Outcome	Probability of Outcome	Batter Goes to	Player on First Goes to	Player on Second Goes to	Player on Third Goes to
Single	0.15	1	2	3	H
Double	0.07	2	3	H	H
Triple	0.05	3	H	H	H
Home run	0.03	H	H	H	H
Base on balls	0.10	1	2	3 (if forced)	H (if forced)
Strike out	0.30	Out	1	2	3
Fly out	0.10	Out	1	2	H (if less than 2 outs)
Ground out	0.10	Out	2	3	H (if less than 2 outs)
Double play	0.10	Out	The player nearest first is out.		

The interpretation of these outcomes is not described in detail. For instance, if there are no men on base, then hitting into a double play is counted simply as making an out.

2. Manager tells player at bat to bunt.

Outcome	Probability	Effect
Single	0.05	Runners advance one base.
Sacrifice	0.60	Batter out; runners advance one base.
Fielder's choice	0.20	Batter safe; runner nearest to making run is out, other runners stay put unless forced.
Strike or foul out	0.10	Batter out; runners do not advance.
Double play	0.05	Batter and player nearest first are out.

3. Manager tells player on first to steal second.

4. Manager tells player on second to steal third.

In either case, the attempt is successful with probability 0.4, the player's position is unchanged with probability 0.2, and the player is out with probability 0.4.

5. Manager tells player on third to steal home.

The outcomes are the same as those above, but the corresponding probabilities are 0.2, 0.1, and 0.7.

Baseball fans please note: No claim is made for the validity of either assumptions or data.

Table II shows the transition probabilities and rewards for a typical state, 0011 or 4. In state $4(i = 4)$, three alternatives apply: hit, bunt, and steal third. Only nonzero p_{ij}^k are listed. The highest immediate reward in this state would be obtained by following alternative 1, hit.

Table II. Probabilities and Rewards for State 4
(0011)

First alternative: Hit, k = 1.

Next State	j	p_{4j}^1	r_{4j}^1	
0000	1	0.03	3	
0100	5	0.05	2	
0110	7	0.07	1	
0111	8	0.25	0	$q_4^1 = 0.26$
1011	12	0.40	0	
1110	15	0.10	0	
2010	19	0.10	0	

Second alternative: Bunt, k = 2.

Next State	j	p_{4j}^2	r_{4j}^2	
0111	8	0.05	0	
1011	12	0.30	0	$q_4^2 = 0$
1110	15	0.60	0	
2010	19	0.05	0	

Third alternative: Steal third, k = 3.

Next State	j	p_{4j}^3	r_{4j}^3	
0011	4	0.20	0	
0101	6	0.40	0	$q_4^3 = 0$
1001	10	0.40	0	

Table III shows for each state i the state description, the alternatives open to the manager in that state, and q_i^k, the expected immediate reward (in runs) from following alternative k in state i. The final column shows the policy that would be obtained by maximizing expected immediate reward in each state. This policy is to bunt in states 5, 6, 13, and 14, and to hit in all others. States 5, 6, 13 and 14 may be described as those states with a player on third, none on second, and with less than two outs.

Obviously, however, the manager should be more interested in the long-run reward--that is, the expected number of runs in the half-inning. The data below were used as an input to a computer program implementing the policy iteration method for determining the policy with the highest expected number of runs in each state [Howard, 1960, Appendix]. Since the program chooses an initial policy by maximizing expected immediate reward, the initial policy was the one just mentioned. The computer had to iterate only twice to reach a solution. Its results are summarized in Table IV.

The optimal policy is to hit in every state. The v_i may be interpreted as the expected number of runs that will be made if the inning is now in state i and it is played until three outs are incurred. Since a team starts each inning in state 1, or "no outs, no men on," then v_1 may be interpreted as the expected number of runs per inning under the given policy.

Table III.   Summary of Input

State Description				Alternative 1		Alternative 2		Alternative 3		Alternative 4		No. of Alternatives in State i	Initial Policy
		Bases		q_i^1		q_i^2		q_i^3		q_i^4			
i	Outs	3 2	1										
1	0	0 0	0	0.03	Hit	—		—		—		1	1
2	0	0 0	1	0.11	Hit	0	Bunt	0	Steal 2	—		3	1
3	0	0 1	0	0.18	Hit	0	Bunt	0	Steal 3	—		3	1
4	0	0 1	1	0.26	Hit	0	Bunt	0	Steal 3	—		3	1
5	0	1 0	0	0.53	Hit	0.65	Bunt	0.20	Steal H	—		3	2
6	0	1 0	1	0.61	Hit	0.65	Bunt	0	Steal 2	0	Steal H	4	2
7	0	1 1	0	0.68	Hit	0.65	Bunt	0.20	Steal H	—		3	1
8	0	1 1	1	0.86	Hit	0.65	Bunt	0.20	Steal H	—		3	1
9	1	0 0	0	0.03	Hit	—		—		—		1	1
10	1	0 0	1	0.11	Hit	0	Bunt	0	Steal 2	—		3	1
11	1	0 1	0	0.18	Hit	0	Bunt	0	Steal 3	—		3	1
12	1	0 1	1	0.26	Hit	0	Bunt	0	Steal 3	—		3	1
13	1	1 0	0	0.53	Hit	0.65	Bunt	0.20	Steal H	—		3	2
14	1	1 0	1	0.61	Hit	0.65	Bunt	0	Steal 2	0.20	Steal H	4	2
15	1	1 1	0	0.68	Hit	0.65	Bunt	0.20	Steal H	—		3	1
16	1	1 1	1	0.86	Hit	0.65	Bunt	0.20	Steal H	—		3	1
17	2	0 0	0	0.03	Hit	—		—		—		1	1
18	2	0 0	1	0.11	Hit	0	Bunt	0	Steal 2	—		3	1
19	2	0 1	0	0.18	Hit	0	Bunt	0	Steal 3	—		3	1
20	2	0 1	1	0.26	Hit	0	Bunt	0	Steal 3	—		3	1
21	2	1 0	0	0.33	Hit	0.05	Bunt	0.20	Steal H	—		3	1
22	2	1 0	1	0.41	Hit	0.05	Bunt	0	Steal 2	0.20	Steal H	4	1
23	2	1 1	0	0.48	Hit	0.05	Bunt	0.20	Steal H	—		3	1
24	2	1 1	1	0.66	Hit	0.05	Bunt	0.20	Steal H	—		3	1
25	3	— —	—	0	Trapped	—		—		—		1	1

The initial policy yields 0.75 for v_1, whereas the optimal policy yields 0.81. In other words the team will earn about 0.06 more runs per inning on the average if it uses the optimal policy rather than the policy that maximizes expected immediate reward.

The values v_i can be used in comparing the usefulness of states. For example, under either policy the manager would rather be in a position with two men out and bases loaded than be starting a new inning (compare v_{24} with v_1). However, he would rather start a new inning than have two men out and men on second and third (compare v_{23} with v_1). Many other interesting comparisons can be made. Under the optimal policy, having no men out and a player on first is just about as valuable a position as having one man out and players on first and second (compare v_2 with v_{12}). It is interesting to see how the preceding comparisons compare with our intuitive notions of the relative values of baseball positions.

Table IV.   Summary of Solution

	Iteration 1				Iteration 2		
State	Description	Decision	Value v_i	State	Description	Decision	Value v_i
1	0000	Hit	0.75	1	0000	Hit	0.81
2	0001	Hit	1.08	2	0001	Hit	1.25
3	0010	Hit	1.18	3	0010	Hit	1.35
4	0011	Hit	1.82	4	0011	Hit	1.89
5	0100	Bunt	1.18	5	0100	Hit	1.56
6	0101	Bunt	1.56	6	0101	Hit	2.07
7	0110	Hit	2.00	7	0110	Hit	2.17
8	0111	Hit	2.67	8	0111	Hit	2.74
9	1000	Hit	0.43	9	1000	Hit	0.46
10	1001	Hit	0.75	10	1001	Hit	0.77
11	1010	Hit	0.79	11	1010	Hit	0.86
12	1011	Hit	1.21	12	1011	Hit	1.23
13	1100	Bunt	0.88	13	1100	Hit	1.11
14	1101	Bunt	1.10	14	1101	Hit	1.44
15	1110	Hit	1.46	15	1110	Hit	1.53
16	1111	Hit	1.93	16	1111	Hit	1.95
17	2000	Hit	0.17	17	2000	Hit	0.17
18	2001	Hit	0.34	18	2001	Hit	0.34
19	2010	Hit	0.40	19	2010	Hit	0.40
20	2011	Hit	0.59	20	2011	Hit	0.59
21	2100	Hit	0.51	21	2100	Hit	0.51
22	2101	Hit	0.68	22	2101	Hit	0.68
23	2110	Hit	0.74	23	2110	Hit	0.74
24	2111	Hit	0.99	24	2111	Hit	0.99
25	3000	Hit	0	25	3000	Hit	0

Epilog

Developments in the theory of Markov decision processes over the past several years have increased their potential in modeling sports.   For example, football or hockey strategists must often consider the time consumed by a play, a consideration of little importance in baseball.   The semi-Markov decision process [Howard, 1971], which explicitly allows possibly uncertain times for state transitions, offers a promising model for such games.

THE VALUE OF FIELD POSITION

by

Virgil Carter and Robert E. Machol

(Based on an article in Operations Research, 1971)

Quantitative evaluation of a strategy in any competitive sport requires a metric on the value of the states which may be reached through alternative strategies. This paper describes a preliminary attempt to construct such a metric for football. Specifically, we have computed the expected value of possession of the football, first and ten, at any point on the playing field, and we have applied these results to some strategic and evaluative considerations.

The expected value of first and ten with n yards to the opponent's goal line is given by $E(X|n) = \Sigma X_i P(X_i)$, $i = 1,...,103$. Four of the outcomes are scores: touchdown ($X_1 = 7$); field goal ($X_2 = 3$); safety ($X_3 = -2$), and an immediate opponent's touchdown following a turnover ($X_4 = -7$). The remaining possible outcomes consist of turning over the ball to the opponents at one of the 99 possible points n where the value is $-E(X|100-n)$. This leads to a system of 99 equations in 99 unknowns. Because of the paucity of data, the field was divided into ten strips: 99 to 91 yards to go, 90 to 81, etc. These data sets are identified by their midpoints in Table I. This condensation led to a system of ten equations in ten unknowns.

Based on a census of the 8373 plays from the first 56 games (7 weeks) of the 1969 NFL schedule, the total number of first-and-ten plays was 2852; the largest data set, centering on 75 yards to go (and including the touchbacks) had 601 points, and the smallest, centering on 95 yards to go, had 57. A separate calculation was performed on the 1258 first downs immediately following possession, and the average absolute difference was less than a quarter of a point. While psychological factors may have affected this difference, it is a reasonable fluctuation for a sample of this size.

The numbers in Table I, which are given to three decimal places, are obviously not reliable to better than one or two tenths of a point, but except for the 5-yards-to-go point they show a nice fit to a line running from about -1.64 points at one's own goal line to +5.62 at the other. The difference between -1.64 and -2 means that, even with the ball on one's own one-inch line, there is still hope; and the difference between +5.62 and +7 means that the last few yards are the hardest. Other than that, the pay-off is quite linear, about 0.0726 points per yard; at the end it increases nonlinearly to 0.2 points per yard or more.

The analysis shows that the expected value of first and ten on one's own twenty yard line (after a touchback) is very close to zero, which indicates wisdom on the part of the rules makers. To check this out more exactly, the plays from the 20-yard line should have been separated out, leading to eleven equations in eleven unknowns. Since the analysis must be done over if the results are to be useful under the new rules, it would also be desirable to make other improvements. The touchdown is actually

Table I. The Expected Point Values of Possession of the Football
With First Down and Ten Yards to Go for Various Ten-Yard Strips

Center of the ten-yard strip (yards from the target goal line): n	Expected point value: $E(X\|n)$
95	-1.245
85	-0.637
75	+0.236
65	0.923
55	1.538
45	2.392
35	3.167
25	3.681
15	4.572
5	6.041

worth less than 7 points, by about .05 because of the possibility that
the extra point will be missed, and by perhaps .5 because of the negative
expected value of the ensuing kick-off (with the new 35-yard-line rule).
The free kick from the 20-yard line has even greater negative expected
value, and the true value of a safety is probably close to -4 points;
however, the safety is so rare that this will probably not greatly affect
the overall analysis.

As an example of the kinds of implications of these numbers, consider
the conventional wisdom which asserts that it is terribly risky to attempt
a pass when one is deep in one's own territory. To check this out, consider a turnover due to an intercepted pass on first down which results
in the opponent taking over, first-and-ten, at the same point on the field.
Then if the turnover occurs at either 45-yard line, the change in value
is (from Table I) 1.538 + 2.392 = 3.930 points; and if it occurs at either
15-yard line it is 4.572 - 0.637 = 3.885 points--values which are clearly
indistinguishable within the accuracy of this analysis. Thus the cost of
a turnover is independent of field position except within the shadow of
either goalpost, where it is slightly greater due to the near-goal non-
linearity. Of course on second or third down the cost of a turnover (and
therefore the risk of a play such as a flare pass which risks a turnover)
is less than on first down, although the exact value cannot be determined
from the present data.

Another significant implication of these data is that the field
goal should probably not be attempted with fourth-and-goal on the two or
three yard line. The expected value of the kick is surely less than 3
points, because it may be missed, and because the ensuing kick-off has
negative value. If one rushes instead, there is a probability, p, of
scoring 7 points; and if one fails to score, one has achieved about 1.5
points by putting the opponent "in the hole". Solving $7p + 1.5(1 - p) = 3$,
we find that the rush should be given preference even if it has slightly
less than one chance in three of scoring.

[In our original paper we added "A similar analysis is pertinent in
evaluating the 'coffin-corner' punt". This tactic was quite rare in
1969, but is now comparatively common; this difference is probably due

to the rule changes concerning missed field goal attempts and punt runbacks, but perhaps was affected by publication of our article.]

Clearly this type of analysis cannot be extended to the determination of the value of second or third down because that will depend not only on field position but also on number of yards to go. However, it could be extended to fourth-and-long situations, where the punt or field-goal attempt almost always follows and the value is therefore virtually independent of exactly how long is "long". The data set will be much smaller since the average number of first downs per series is 2852/1258 = 2.27, while the number of fourth-and-longs is slightly less than one per series. However, with this data, one could evaluate the individual teams (offense, defense, and specialty). For example, suppose a team received the ball, first and ten, on its own 45 and then reached fourth and long after advancing to the opponent's 35. The offense of that team would then, during its possession, have changed its expected value from +1.538 points to whatever the other situation might turn out to be -- say +0.637 points (if one might be expected to kick out of bounds, on the average, at the 15-yard line). This offense would then have lost 0.901 points, and the opposing defense would have won 0.901 points. One could similarly evaluate each series, from first possession to score, turnover, or kick, as well as the kicks themselves, to find the exact contribution of each of the three teams.

We would like to take this opportunity to suggest one other area in which most professional football teams seem to be making an incorrect choice of strategy. No statistics have been gathered to support this, but the error (as we assume it to be) does derive from a naive misunderstanding of decision theory. We refer to the strategy of calling time-outs during the last two minutes of either half.

If one is considering calling a time out, the Type I error is in calling it when one should not, and the penalty arises when the ball is turned over to the opponents before the time expires. This may happen through a turnover, through running out of downs followed by a punt, through a missed field-goal attempt, or through an actual score after which the opponents again get the ball. This Type I error appears to be a very frequent occurrence. As a well-known example, we cite the Cleveland-Oakland game of November 8, 1970, in which Cleveland, attempting to break a tie, called a time out and subsequently lost the ball by an interception; then Blanda had enough time to win the game by kicking a field goal.

The Type II error consists of not calling a time out when one should. The naive assumption appears to have been that the penalty for this is running out of time. This is wrong. The penalty for a Type II error is running out of time when one still has time outs left to call, and we believe this to be an extremely rare occurrence.

That is, we assert that the Type I error, calling a time out when one should not, and being penalized for it, has been made many times; whereas, the Type II error, failing to call a time out when one should, is extremely rare. Specifically, therefore, we recommend that a team should never call a time out for the exclusive purpose of stopping the clock when there are more than 30 seconds to play, if it has the ball.

COLLEGIATE FOOTBALL SCORES, U.S.A.

by

Frederick Mosteller

Based on an article in
Journal of the American Statistical Association, 1970

Scoring in Football. The scoring in 1967 collegiate football allowed
six points for a touchdown with a bonus opportunity to try for a one or
two-point conversion, one point for kicking a point after touchdown, two
points for a running-play conversion after touchdown (the kick being the
more frequently chosen), three points for a field goal, and two points for
a safety, a rare play. Since touchdowns followed by successful one-point
conversions are frequent, we can expect many scores to come in multiples of
seven or nearly that. Field goals, while frequent, were not as frequent as
in professional football, partly because the goal posts were ten yards farther away in the collegiate game.

Because of the different number of points awarded for the varying
scoring plays, the scores in football may carry more information than their
mere numerical value. For example, we shall see that a final losing score
of three is very much more than three points better than a losing score of
zero.

During the years since collegiate football began, there have been many
changes in the playing rules, and these have affected the scoring. Since
this analysis deals primarily with 1967 scores plus a few comparisons with
1968 scores, these changes would need to be considered for comparisons with
earlier or later years. The 1967 and 1968 seasons were played under essentially the same rules.

Frequency Distribution--Winner's Score versus Loser's. To begin, let
us look at a frequency distribution of the scores of these 1158 games.
Since each game produces two scores, a two-way table offers a convenient
summary of all the outcomes. Table I shows the number of times each pair
of scores occurred, with winner's score along the top and loser's down the
side. For example, in six games the winner had three and the loser zero
points.

Ties are shown along the diagonal extending southeast from the upper
left-hand corner. The "3" in the upper left-hand corner means that three
games among these 1158 ended in 0-0 ties during 1967. (We will usually
speak of winning and losing scores with the understanding that this language includes the two scores involved in a tie.) In earlier times, 0-0
were more common. To check this, Mrs. Holly Grano tabulated the 0-0 ties
among the scores of earlier collegiate games and found 1.5 percent 0-0 ties
in 196 mainly Eastern games in 1920; 3.3 percent 0-0 ties in 585 games in
1925; 3.8 percent 0-0 ties in 765 games in 1930; and 2.8 percent 0-0 ties
in 564 games in 1935, compared with 1967's 0.3 percent. In 1967, the tie
with the most scoring was 37-37 for Alabama versus Florida State.

Table I gives us a way of assessing the rarity of the Harvard-Yale
29-29 tie in 1968. First, ties themselves are rare these days, only 26 in

Table 1. Frequency Distribution of 1967 College Football Scores[a]

Losing score	Winning score 0 2 3 4	6 7 8 9	10 11 12 13	14 15 16 17	18 19 20 21	22 23 24 25	26 27 28 29	30 31 32 33	34 35 36 37	38 39 40 41	42 43 44 45	46 47 48 49	50 51 52 53	54 55 56 58	60 61 62 63	65 67 68 69	70 75 77 81 90	Totals
0	3 6	4 14 1 3	4 3 5	12 3 7 12	5 10 11	2 4 4 1	5 4 9 5	3 6 3	7 11 1 4	5 1 5	4 2 2	4 2 1	4 1	2 3 2	1 1 1	1 1	2 1 1	219
2	1	1 1		1	1		1			1								6
3		1 5 1	1 3	4 1 1 1	1 , 3 2	1 1	1 1 1		1 1 1	1 1		1						36
4									1									1
6		1 11 1 1	1 5	6 1 3 5	2 1 5	1 4 6	1 4 8	4 3	5 3 2 2	2 4	4 2 1	1 1 1	3 3 1	3	1		1	114
7		3 1 7	12 1 5 8	16 1 2 9	3 4 10 14	3 7 8 1	2 6 10 3	8 3 2	2 3 2	2 1 , 6	1 1 1	2 3 2 1	1	2 1 1 1	2	1 1		186
8		2 2 4	1 1	2 2	1 1 3 1	1 1 1	1 1	1	2 1 1		1		1			1		32
9	1	3 1	1	1 1	2 1	2 1	2 1	1		2							19
10		1 2 4	10 3 2 2	1 4 6	1	2 2	3 1	1	1	2							48
11		1			1		1		1								4
12		5	7 1	1 1 1 4	1 2 3	2 1	1 1	1 2 1	1	1		1	1 1		1 1		42
13		4	12 3 2 7	3 1 2 4	2 3	3 4 3 1	3 2	1	2	1 2 2 1	1						72
14			4 3 4 8	1 4 10 12	1 7 7 1	4 7 11 1	2 3 2	6 6 1 1	2 1 2	5 1 2 1	2 1	1 3	1				128
15			1 3	1 4	1 1 2	1	1 1 1		1 2	1		1	1				23
16			3 3	2 4 4	2 1 1	2 1	1	2	1 1	1		1					29
17				1 5	4	1 2 2	1 2 1	1 1 1	1	1							23
18				2 2 2	1	1 1	2	1	2								14
19				2 3	1 1	1 1	1 1	1	1								14
20				5	1 1 4	1 4 1 3	2 2	1	1	. 1	1 1						29
21	.			1	4 3 2	2 3 5 1 1	2	1 4	1 2 2	1	1	1					36
22					1 2 1	1	1 1 1	1	1 1 1	1							14
23					1 2		2		1	1	1	1					8
24					1	2 1 3	1 1 1	· .	1 1	1							13
25							2	2 1	1								5
26						1		1		.							2
27						2	1	2	2 1 , 1 1								10
28						1	2 3 1	1	1	1							11
29							1 1 1	1									4
30							1 1										2
31							1 1	1									3
32							1		1	1							3
33								1									1
34								1	1								2
35								1 1									2
36																	0
37								1	1								2
38									1								1
Totals	3 0 7 0	7 34 3 13	24 1 13 39	74 15 27 52	12 22 55 82	20 35 53 8	27 42 61 18	30 35 9 15	30 45 9 20	18 10 7 28	30 12 7 7	6 13 12 7	6 5 8 2	4 3 10 4	1 4 1 1	1 1 1 2	1 1 3 1 1	1158

[a] Scores taken from [Long, 1968].

all or two percent. Only three games had ties with higher scores. Looked
at another way, only 20 games produced scores of at least 29 by both teams.

Frequent Scores. What are the popular scores for a game? The most
popular score for a game as shown in Table I is 14-7, which occurs 16 times,
followed closely by 7-0, 21-7, with 14 occurrences each, then 10-7, 14-13,
17-0, 21-14, with 12 each. Among the higher scoring games, the most frequent are 28-7, 28-14, and 35-0, with frequencies of 10, 11, and 11, respectively. As we anticipated, multiples of seven are prominent among these
popular scores.

The team scores that occur with high frequency usually stand out from
those near them because of the lumpiness of the scoring system. We can see
this even more clearly in the marginal totals of Table I. A special tabulation of these in Table II shows the frequencies of those scores. The
total counts with asterisks show the six most frequently occurring scores.
These six most frequent team scores account for 994 scores in all, which
represent about 43 percent of all the team scores in our 1158 games. These
frequent scores are 0, 7, 14, 6, 21, 13 in descending order of frequency.
The first three are nearly twice as popular as the last three. The six
most frequent winning scores are 21, 14, 28, 20, 24, 17 in descending frequency. The most frequent losing scores are 0, 7, 14, 6, 13, 10 in descending frequency.

Note that the single most frequent winning score is 21; the most frequent losing score is 0, but the most frequent game score is 14-7, whose
components do not match either of these. Of course, 21-0 was frequent,
and the difference in count between its 11 and the 16 for 14-7 is not so
large that it could not be reversed in data from another season.

Table II

Distributions of Team Scores up to Scores of 29

Score	Winning	Losing	Total	Score	Winning	Losing	Total
0	3	219*	222*	17	52*	23	75
2	0	6	6	18	12	14	26
3	7	36	43	19	22	14	36
4	0	1	1	20	55*	29	84
6	7	114*	121*	21	82*	36	118*
7	34	186*	220*	22	20	14	34
8	3	32	35	23	35	8	43
9	13	19	32	24	53*	13	66
10	24	48*	72	25	8	5	13
11	1	4	5	26	27	2	29
12	13	42	55	27	42	10	52
13	39	72*	111*	28	61*	11	72
14	74*	128*	202*	29	18	4	22
15	15	23	38				
16	27	29	56				
				Total	1158	1158	2316

* One of six most frequent scores in the column.

Figure 1. Proportion of Times a Given Score Wins

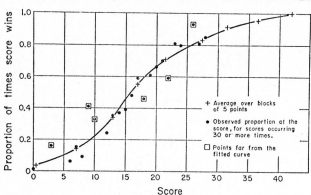

The average winning score is 26.7, the average losing score 10.1, for an average difference of 16.7 points. And so in round numbers the average game score is 27-10, as opposed to the most frequent score which is 14-7. The median winning score is 24, the median losing score is 8, and so 24-8 might be called a "median" or middling game.

The Chance that a Score Wins. How often does a given score, say 10, win? Are some scores relatively good and some poor considering their size? We now look at each team score and ask to what extent it is a winning score. Table II shows the basic data. Figure 1 displays the information more clearly. Ties are counted as 1/2 for each team.

The lazy S-shaped curve in Figure 1 shows the general rise in probability of winning with increasing score. It is a freehand curve that passes through the +'s which are based on average results for 5 scores (the intervals 0-4, 5-9, 10-14, etc.). We see that a score of 16 has a 50-50 chance of being a winner. A team with 40 points is practically assured of victory.

The dots which correspond to single scores generally fall close to the curve. The six boxed dots furthest from the curve are instructive. For example, a score of 9, which the average curve predicts will win only 18 percent of the time actually wins 41 percent of the time, and thus it is a very special score. The scores 3, 9, 10, are favorable scores. That is, teams with these scores are considerably more likely to win than the size of the score alone indicates, an additional 9 percent for 3, 22 percent for 9, 11 percent for 10. These are all scores intimately related to field goals. The score 26 seems also to be especially good, 12 percent above the curve, though perhaps it is due to sampling fluctuations.

The unlucky scores are 18 and 22, scoring 14 percent and 13 percent below the average for scores near this size. Of course, Notre Dame once won a famous game with 18 points, but that score does generally mean missing 3 points after touchdown, rather a bad sign. Probably 22 points comes from three touchdowns, one successful and two unsuccessful one-point conversions, and one field goal. But it could arise in many other ways including two successful one-point conversions and one successful two-point conversion. The proper interpretation awaits a further breakdown of the methods of scoring by which the 22 points are achieved, but which our data cannot offer.

Predicting Winner's Score From Loser's. How does the winner's score change as the loser's increases? To relate the winner's score to the loser's, let us compute the mean winning score for each losing score as long as there are enough games to make the calculation reasonably reliable. Table III shows the results. Its first row gives 0 as the loser's score in a total of 219 games, and an average winning score of about 27 points. The corresponding calculations are made for each of the rows having at least 10 games. Results for rows 23 to 33 inclusive are pooled and plotted in Figure 2 at 28 for losers and 34.7 for the winners. These numbers give predictions of winning scores for each losing score.

We have drawn a smooth curve through these points to represent an idealistic smooth forecast of winner's score from loser's score, by groups of 5 losing scores. This curve is roughly hyperbolic in shape. This curve is idealistic in that it ignores any special information contained in the scores beyond their numerical values. By looking at the departures of the points from the curve, we can tell what some of this information is. In particular, a score of 0 by the loser means that the team is scored on more frequently than the fitted curve would indicate--that is, that the average team with 0 score is more scored upon than the mere zeroness of their score might lead one to expect. This idea is even clearer when we move up to a score of three points for the loser.

Note that for a loser's score of 3, the mean winning score falls well below the smooth fitted curve. This suggests that a team that has scored one field goal can be thought on the average to be showing additional competence (relative to the other team, it may of course just be better matchmaking) in defense--about 6½ points worth. On the other hand, a score of 6, usually reflecting a touchdown without extra point, is a slight indication of less competence.

No doubt the implication is two-fold, on the one hand not being able to kick the extra point is itself a sign of less competence, and on the other hand having a kick blocked is a sign of extra competence on the part of the other team. And so either way one looks at it, a score of 6 by the loser suggests better scores for the winner than the smooth curve would predict.

One might then expect that a score of 7 on the part of the loser would make them look a little better, and relative to a score of 6 they do perform better, but they fall just about on the smooth curve. A losing score of 9 or of 10 shows again a good deal of strength or that the teams are more evenly matched than the smooth curve implies, again perhaps because of the field goal, or because the field goal was worth trying for.

Note that the curve of Figure 2 generally slopes upward. The reason is that the more the losing team scores, the more the winning team must score to keep ahead. And, of course, the curve has to lie entirely above the 45° line through the origin.

Table III.

Average Winning Score for Each Losing Score

Losing Score	Number of Games	Average Winning Score	Losing Score	Number of Games	Average Winning Score
0	219	26.7	20	29	28.9
2	6	19.7	21	36	30.8
3	36	19.0	22	14	33.6
4	1	35.0	23	8	36.1
6	114	27.8	24	13	31.3
7	186	24.5	25	5	34.8
8	32	25.1	26	2	31.0
9	19	21.8	27	10	37.4
10	48	20.4	28	11	34.7
11	4	28.5	29	4	33.8
12	42	27.0	30	2	33.0
13	72	24.3	31	3	34.3
14	128	27.8	32	3	41.7
15	23	27.4	33	1	36.0
16	29	24.6	34	2	38.5
17	23	29.2	35	2	36.0
18	14	30.6	37	2	39.5
19	14	29.6	38	1	40.0
			Total	1158	26.7

Differences in Score. Next, let us look at the frequency distribution of the differences between the scores. This is shown in Figure 3. The horizontal axis of Figure 3 shows the difference d between the winner's and the loser's score, the vertical axis the frequency of that difference in games. Within each block of 10 scores (0-9, 10-19, etc.) we put a + at the median--the number midway between the two middle numbers in the block, both horizontally and vertically. Then a smooth curve was drawn through these medians. Some counts which stand especially far from the curve either above or below have been boxed. These numbers again are ones that have some special significance over and above the relation implied by the curve. For example, a difference of 10 seems to occur too seldom if the fitted curve is to be believed. Why would this happen?

To understand this, it helps to think of close scores and named teams instead of winners and losers. As every dice player knows, 5-5 is "the hard way" to make 10, and is only half as likely as 4-6. Something like this happens in football. When Harvard and Yale play, a score 14-13 can happen two ways, but a score of 14-14 can happen only one way. We would guess that scores very close to one another would be nearly equally frequent, except for lumpiness. Thus if we think of teams A and B as nearly equally matched and of the scores being assigned to them in the order A, B, then we think of 14-13, 14-14, and 13-14 forming three nearly equally likely game scores. Then we should find that the ties happen about half as often as the one-point differences. Some may worry that the 14-13, 14-14, 13-14 idea neglects the possibility of a 13-13 tie also. But to check this out, one need only think of a large-size checkerboard as in Figure 4 to see that there are two diagonals, one above and one below the main diagonal, and each essentially as long as the main one.

Figure 2. Graph of Average Winning Score for Each Losing Score[*]

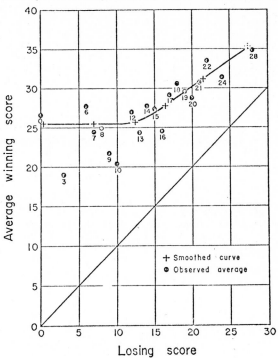

Figure 3. Frequency Distribution of Winner's
Score Minus Loser's Score

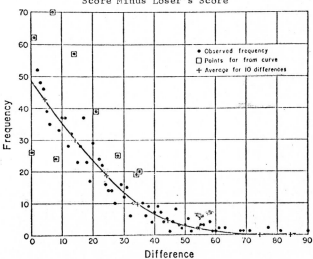

This argument makes it reasonable that ties should happen only about half as often as scores differing by one point, or even by two points. Consequently, we should look to see whether the number of differences of 1 is about twice the number of differences of 0. There are 62 differences of 1 and 26 differences of 0, so when we double the ties, we are still left with an excess of 10 differences of 1. This difference is probably due to the fact that a team which has arrived at a one-point deficit by scoring a touchdown late in a game usually tries a two-point play, resulting in winning or losing by one point, rather than electing to kick for a tie; see [Porter, 1967].

Aside from those of 0 and 1, we note that the frequencies of differences of 7, 14, 21, 28, 34, and 35 lie well above the curve, presumably reflecting the 7-pointness of the touchdown plus kicked point after. Points falling below the curve are a bit harder to identify, but 8, 19, 27, and 32, seem to fall below, although only 8 is especially convincing. One might have thought that 14-6 would be a frequent score that would produce many 8's, but that did not happen.

When we plotted the sum of the scores against the difference of the scores, and computed the average sum against the average difference over a short interval, we found the resulting points to be fitted well by the hyperbola shown in Figure 5 whose equation is

$$s = \sqrt{900 + d^2}$$

Figure 4.
Checkerboard Layout Suggestions that Ties are
About Half as Frequent as Differences of One Point

Team A's score

		0	1	2	3	4	5	...
Team B's score	0	tie	d=1					
	1	d=1	tie	d=1				
	2		d=1	tie	d=1			
	3			d=1	tie	d=1		
	4				d=1	tie	d=1	
	5					d=1	tie	
	.							

We have no theory to suggest this curve. It is true that we expect the sum and difference to become identical as the difference grows large. This suggests the line having equation s = d as asymptote. The curve is close in general shape to that of average winner's score for given loser's score (Figure 2). The fact that even when the difference is as large as 60 the curve has not gotten close to its asymptote suggests that even a weak losing team will score a touchdown on the average. In fact, Table I shows that both teams will get at least six points in more than three out of four games.

We noted earlier that the average difference between scores is 16.7. We wondered how this average difference would compare with the difference one would get by choosing two team scores at random from all the 2 x 1158 = 2316 scores made by winning and losing teams. Thus we simulated a random game. The calculation gave an average difference of 14.7. Thus randomly assembled team scores would be more closely matched than the scores that actually occurred. Random pairing of scores is equivalent to regarding all teams as equally skilled, and so we should expect a smaller mean difference, because in real games the teams frequently are badly mismatched. To illustrate, the number of 0-0 ties was 3, but under random pairing there would have been 10 scoreless ties.

Figure 5.   The Average Sum of Scores For a Given Difference With a Hyperbola Fitted to the Data

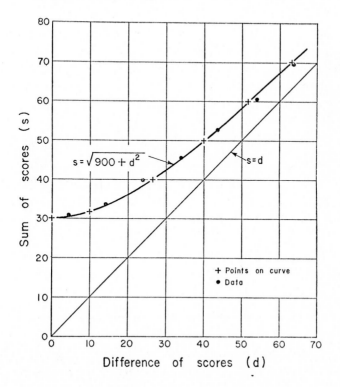

ANALYSIS OF SCORES OF IVY LEAGUE FOOTBALL GAMES

by

Shelby J. Haberman

Attempts to predict results of football games are frequently made by gamblers, spectators, and the mass media. Success often appears limited; however, this limited predictability appears to reflect inherent variability in results of football games rather than failure to develop statistical models of sufficient sophistication. To investigate this claim, an analysis was made of final scores and box-score statistics from the 196 Ivy League football games played from 1960 through 1966. The choice of the Ivy League partly reflects the fact that the analysis described was made when the author was at Princeton; however, the Ivy League also has the attraction that each of the eight teams in the league plays every other team once in a season.

To obtain an initial assessment of the predictability of games, a three-way analysis of variance was performed on the final scores of all games played between Ivy League teams from 1960 to 1966. The analysis-of-variance model can be written in the following manner.

$$S_{ijt} = m + I_i + J_j + T_t + a_{ij} + b_{it} + c_{jt} + \varepsilon_{ijt}$$

Here S_{ijt} is the score of team i, $1 \leqq i \leqq 8$, against team j, $j \neq i$, in year t, $1 \leqq t \leqq 7$; m is an average score; I_i might be how good Princeton is on the average and b_{it} how much better in year t; T_t might represent a particularly high-scoring year; a_{ij} the special factors affecting a Harvard-Princeton game; and ε_{ijt} is the random component. Customary constraints on parameters are employed so that

$$\Sigma I_i = \Sigma J_j = \Sigma T_t = \Sigma a_{ij} = \Sigma a_{ij} = \Sigma b_{it} = \Sigma b_{it} = \Sigma c_{jt} = \Sigma c_{jt} = 0.$$
$$_{i j t i j i t j t}$$

The error terms ε_{ijt} are assumed independently distributed with variance σ^2_{ijt}. The assumption of independence of ε_{ijt} is not obviously valid, especially since ε_{ijt} and ε_{jit} involve scores from the same game; however, an investigation of possible correlations of scores suggests that the independence assumption is tenable.

Given the proposed analysis-of-variance model, one obtains a residual mean square s^2 of 96.5. This mean square is an unbiased estimate of the average variance

$$\sigma^2 = \frac{1}{392} \sum_{t=1}^{7} \sum_{i=1}^{8} \sum_{j \neq i} \sigma^2_{ijt}$$

of the ε_{ijt}. This result suggests that standard deviations of scores are on the order of 10 points. Standard deviations σ_{ijt} appear related to the expected value of S_{ijt}, with largest values of σ_{ijt} when expected values are between 20 and 40 points.

This analysis of variance suggests that scores are very difficult to predict. The residual mean square of 96.5 should be compared to the total mean square

$$\frac{1}{391} \underset{t}{\Sigma} \underset{i}{\Sigma} \underset{j \neq i}{\Sigma} (S_{ijt} - S...)^2$$

of 173. Thus if predictions of scores were made with full knowledge of the unknown parameters m, T_t, I_i, J_j, b_{it}, c_{jt}, and a_{ij}, the estimated value of σ, the square root of the expected mean-square error of prediction, would be 9.8. Given the best constant prediction of the score S_{ijt} for all teams i and j and years t, the estimated square root of the expected mean-square error of prediction would be 14.1. Since, in reality, parameter estimation is required for predictions to be possible, the gain in accuracy over the constant predictor is likely to be remarkably limited.

Even prediction of final scores from box-score statistics for the same game is a rather difficult task. A linear-regression analysis was performed with data for games from 1964 through 1966. The dependent variable was the score S_{ijt}. Independent variables included S_{jit}, the other team's score, together with the following box-score statistics for each team: first downs, yards rushing, yards passing, passes completed, passes attempted, interceptions, number of punts, punting average, fumbles lost, and yards penalized. The residual mean square after this regression analysis with 21 independent variables is 45.1. The corresponding standard deviation is 6.7 points, which is about equal to one touchdown.

A relatively small number of box-score statistics do nearly as well in prediction of final scores as do all 21 predicting variables. If the six variables yards rushing, yards passing, passes completed, passes attempted, interceptions by, and fumbles lost by opponent are used, one obtains a residual standard deviation of 7.4. The most important individual predictor of final scores is yards rushing. The R^2 from use of this single predictor is 0.45, while the R^2 from yards rushing and yards passing together is 0.58 and the R^2 for all 21 predictors is 0.79.

Some insight into the regression analysis may be obtained by consideration of the regression coefficients of the six-predictor model. Yards rushing and yards passing both have regression coefficients of about 0.1, so that 10 yards gained corresponds to a predicted gain of a single point or 70 yards corresponds to 7 points. Interceptions by has a regression coefficient of about 2.9, and fumbles lost by opponent has a coefficient of 2.2, so that each turnover corresponds to a predicted gain of 2 or 3 points. Passes completed has a regression coefficient of 0.49, while passes attempted has a coefficient of -0.66. These coefficients are harder to interpret than those for turnovers or yards gained rushing or passing. They appear to reflect the tendency for passing to be increasingly frequent when a team is behind.

Prediction of box-score statistics based on previous box-score statistics appears about as difficult as prediction of final scores. Three-way

analyses of variance were performed on all box-score statistics using the
same procedure employed with final scores. The ratios of residual mean
squares to total mean squares ranged from a low of 0.49 in the case of
yards rushing to a high of 1.1 in the case of fumbles lost. In terms of
square roots of expected mean-square errors of prediction, the ratio of
0.49 corresponds to an estimated 30% reduction. The ratio of 1.1 only
exceeds 1 because of sampling error. The ratio suggests that neither
knowledge of the teams playing nor knowledge of the year assists in prediction of the number of fumbles lost during a game.

In general, Ivy League games appear very unpredictable, both in terms
of final scores and in terms of box-score statistics. The extent to which
these conclusions apply to other football conferences is unclear; however,
the data do suggest that football games are subject to much more random
fluctuation than most observers would anticipate.

EXTRA-POINT STRATEGY IN FOOTBALL

by

Richard C. Porter
Based on an article in The American Statistician, 1967

Scarcely an autumn Saturday passes that does not see at least one college football team scoring a touchdown in the closing minutes to cut its deficit to one point. At the moment of the score, the stands begin to buzz with opinions: should the coach direct his team to seek a tie by means of the relatively sure one-point place-kick or go for a victory by means of the less certain two-point effort (either a run or pass, but hereafter called a "run"). A less common but more interesting (and analytically more complex) predicament occurs when a team finally begins to move in the fourth quarter and scores a touchdown to cut its deficit to eight points (e.g., to 14-6). There is time for another touchdown. In preparing his strategy at this point, the coach should assume that his team will in fact score this one more touchdown--and that no other scores will occur-- since otherwise the extra-point choices are largely irrelevant. The question here is: should the team 1) kick now, planning to kick again after the second touchdown for a tie, 2) kick now, planning to run later for a win, or 3) run now, planning to kick later for a win? Analysis of these two situations -- where the team trails by one point or by eight points -- is sufficient to illustrate the choices involved. What follows is therefore not a general theory of the extra point. For one thing, in neither of these two situations is the possibility of an opponent's later touchdown considered. Extra-point strategy following touchdowns early in a game contains complicating game-theory elements which are neglected here.

The correct ranking of a team's preferences for the various extra-point strategies should conform with its expected-utility ranking of the prospects [Baumol, 1961; Luce and Raiffa, 1957]. Implicit in the decision is knowledge of the utility to the team of a victory, a tie, and a loss in the game, and also estimates of the probabilities that a kick and a run will succeed. Since two points on any utility schedule may be chosen arbitrarily, define the utility of a victory as unity and the utility of a loss as zero. Then, the utility of a tie is equal to t, where $0 \leq t \leq 1$. The probability of a successful two-point effort is assumed equal to r and of a successful kick equal to k. For most teams, k is significantly larger than r, and indeed the strategy choices are trivial if $k < r$.

When a team scores a touchdown late in the game, thereby cutting its deficit to one point, it has two extra-point strategy choices:

A. It may try a two-point play, winning if the attempt succeeds and losing if the attempt fails. The expected utility of such a strategy is:

$$EU_A = r(1) + (1-r)(0) = r \qquad (1)$$

B. It may kick the extra-point, achieving a tie if the kick succeeds and losing if it fails. The expected utility of this strategy is:

$$EU_B = k(t) + (1-k)(0) = kt \qquad (2)$$

Clearly the preference for Strategy A or B should depend upon whether:

$$r \gtreqless kt \qquad\qquad (3)$$

The fact that most teams, in this situation, will play for the win is consistent with inequality (3) either if r is fairly high compared to k or if t is fairly low. Knowledge of all three parameters is critical to the decision. The coaches' maxim, "always play for the win", is not a bad rule-of-thumb approximation of inequality (3) if, for example, t is almost always less than one-third, and r is never as low as one-third (at least for teams capable of scoring at all).

When a team trailing by 14 points late in the game scores a touchdown and has prospects of scoring another, it has three basic strategies available:

I.   It could make a two-point try on the first extra-point. If that succeeds, a kick is planned after the second touchdown (not a run at that time since r is less than k). If the first try fails, the team must run again later in an effort to tie.

II.   It could try a kick after the first touchdown. If that succeeds, another kick is planned in the later effort to tie. If the first kick fails, a run for a tie must be tried later.

III.   It could try a kick after the first touchdown. If that succeeds, a run for the win is planned after the second touchdown. If the first kick fails, a run for the tie must be tried later.

For each of the three strategies the probability of a win, a tie, and a loss can be easily calculated (on the assumptions that a second touchdown will be achieved and that the probabilities r,k, and of scoring again are unaffected); these probabilities are shown in the table below:

		Probability of	
Strategy	Win	Tie	Loss
I.	rk	$r(1-k) + r(1-r)$	$(1-r)^2$
II.	0	$k^2 + r(1-k)$	$k(1-k) + (1-k)(1-r)$
III.	rk	$r(1-k)$	$(1-r)$

The expected utility òf each strategy can then be calculated:

$$EU_I = rk + [r(1-k) + r(1-r)]t \qquad\qquad (4)$$

$$EU_{II} = [k^2 + r(1-k)]t \qquad\qquad (5)$$

$$EU_{III} = rk + [r(1-k)]t \qquad\qquad (6)$$

First, compare Strategies I and III. By inspection it is clear that for any values of k,r, and t (between zero and unity) the expected utility of the first strategy is higher; therefore, Strategy III should never be chosen. Kicking now with the intention of running later for a win can never be optimal strategy. Yet whenever economist-fans are polled on their preferences among the three strategies in this situation, a significant fraction inevitably choose the dominated strategy (and not because they quarrel with the assumptions of the analysis).

The choice between Strategies I and II does not yield so clear-cut a conclusion. Comparison of the expected utilities of the two strategies indicates that Strategy I is superior or inferior to Strategy II according as:

$$rk + [r(1-k) + r(1-r)] \ t \gtreqless [k^2 + r(1-k)]t \qquad (7)$$

that is, according as:

$$t \lesseqgtr \frac{rk}{k^2 - r(1-r)} \qquad (8)$$

Inequality (8) assumes that k^2 is greater than $r(1-r)$; if that is not true, Strategy I is definitely superior to Strategy II. However, this worry is pedantic since $r(1-r)$ cannot exceed .25 and k is generally well above .50.

The choice between Strategies I and II depends importantly upon the values that a team assigns to k,r, and t. Though the likely direction of inequality (8) is not immediately apparent, a few numerical examples indicate that Strategy I will usually be preferred. Let k equal unity, since this biases the numerical examples towards Strategy II (which involves two kicks); then if t is no higher than .25, r need only exceed .21 to make Strategy I superior; if t does not exceed .50, an r of .38 or better is sufficient; and even if t is as high as .75, r need only exceed .57 to make the expected utility of Strategy I higher. Thus a coach who chooses to kick two extra-points in this situation (i.e., seeks the tie) must believe that the probabilities of a successful two-point effort are quite low and/or that the utility of a tie to his team is quite high.

INFORMATIONAL ANALYSIS AND PROFESSIONAL FOOTBALL

by

Frank Ryan, Arthur J. Francia, and Robert H. Strawser
Based on an article in Management Accounting,* 1973

During a National Football League game between the New York Jets and
the Cleveland Browns, in the second quarter the Browns needed seven yards
on third down. Wide receiver Gary Collins raced down the right sideline,
trying to elude the two Jet defenders who were double-covering him, when
suddenly his teammate, tight end Milt Morin, curled over the middle to take
the pass and make the first down with several yards to spare. After the
play had ended, it was obvious to nearly everyone watching that Collins was
in fact a decoy and Morin had been the intended receiver all along.

A thorough analysis of the Browns' tendencies by the Jets could possibly
have nullified the completion. The Browns at the time were primarily a pattern team, executing most of their pass plays according to predetermined
plans, which incorporated their field position, the down, the yardage needed
to gain a first down, the game situation, and anticipated defense. If the
Jets had sufficiently familiarized themselves with the Browns' modus operandi,
they would have known that the Browns, who led by a touchdown at the time,
rarely use the long pass on third down, instead relying on short passes to
the backs and to the tight end. With this knowledge, the Jets could have
played percentage ball by assigning a single man to cover Collins, thus free-
ing the other defensive back to assist in the coverage on Morin.

This situation illustrates a fact of modern football life. The sport
is structured in such a way that careful acquisition and evaluation of the
observable details associated with each play can lead to meaningful trends
analysis. In fact, most coaches, on all levels of the sport, compile some
statistics or frequency counts which convey information on their opponents
as well as themselves. Unfortunately, to accomplish this task manually in
a thorough and complete manner is time consuming to the extent of being impossible in the practical world of winning football.

The application of data processing to professional sports is developing quite rapidly [Purdy, 1971]. In football, computer technology finds an
opportunity to provide a practical and valuable assist to the coaching method
in the area of strategy preparation. While it is standard operating procedure to review the films of previous games and to chart tendencies in a
variety of situations, data-processing methods can nevertheless lead to
inadequate and even misleading information.

To meet the information needs of football coaches, PROBE was conceived**
as a means of utilizing the computer to process and evaluate football data.
It is a generalized report-generating system for application in football
strategy analysis. This concept is based on the fact that football possesses a great number of easily identifiable characteristics, such as down,

---

*Parts of this article reprinted by permission of the National Associa-
tion of Accountants, New York.

**The system was developed by Frank Ryan Computer Service in conjunction
with the Chi Corporation of Cleveland, Ohio.

distance, field position, relative score, offensive formations, defenses, weather, personnel, etc., all of which are directly related to the decision-making processes of the game. These factors must be evaluated relative to one another in order that new information results.

The system is effected as a multistep operation. First, the coaching staff must define in a precise and consistent fashion the complete terminology that it requires to describe the football events that confront the coaches. Terminology varies from team to team, and in fact the basic structure of football ideas changes from coach to coach. For the main part these terminology schemes are incomplete, serving only to describe what the coach's own team does, not what the opponents do. It is not an easy task to establish a precise terminology base which is rich enough to describe all football maneuvers, and which is simple to use in practice.

An input form is prepared which includes all of the basic play terminology required to describe the football factors being analyzed. The design of this input document is critical because of the timing problems involved. Obviously, the more information desired, the more time is required to encode the data. Since the time of the coach or scout encoding the information is limited, it is essential that the input document allow him to gather all of the information needed in the minimum time possible.

The basis of football analysis is the study of game films in which each play is recorded in detail on a separate input form. A professional game will normally include anywhere from 100 to 150 plays, offensive and defensive combined. Films are viewed by a coach (or coaches) and each play is analyzed and recorded. Items such as the following might be included among the encoded information: the number of the play (all plays in the game are numbered sequentially); down; yardage necessary for first down; field position; type of formation; type of play; fakes; backfield action; blocking action; patterns run by receivers; configuration of the defensive line; shifting patterns of the defense; operational defensive actions that are repeated; nature of the pass coverage; result of the play; weather; and personnel.

The information recorded on each input sheet can be transformed in a variety of ways to machine-readable form, thus creating a football data base. Each week data are added to the information bank, updating the data base.

At this point it should be noted that the scouting function is often also performed "live". Usually a coach will send one of his assistants to scout a game which involves a future opponent. While this phase of scouting lacks much of the precision found in extracting data from films, it frequently provides an important supplement to the analysis of game films. For example, it provides information on weather, tempo, mood, sideline signals, etc. It should be noted that while "live" scouting depends on the intuitive grasp or feel of the scout, for many years it was the only method teams had of scouting an opponent.

The overall purpose is, of course, to obtain as much information as possible concerning the specific strengths and weaknesses of an opponent. While each play in a football season is an event that will never specifically recur, the strengths and weaknesses of a team do recur in well defined statistical patterns which are particularly susceptible to trend analysis. Simply put, most football teams normally follow similar patterns of play when confronted with similar situations, modified by predetermined strategies. Thus, for planning purposes in football it is important that detailed information concerning both your opponent and your own team be known to management, i.e., the coaches.

The information desired by the coach should, of course, determine the manner in which the data will be manipulated and also the format of the subsequent report generation. The analysis technique developed must be flexible enough to allow the report to provide answers to the "usual" football analysis questions with minimal effort. It should also be flexible enough to retrieve "unusual" information on command as required. Just as is the case in a business situation, the report should fit the needs of the user, rather than fitting the user to the report.

There is need for a general-purpose "standard" report which shows the overall tendencies; e.g., running game, passing game, and use of personnel. This general report should be sophisticated enough to indicate a profile of the team's operational decisions and tendencies. This type of information depends on the reliability of the football data base. As an example of the type of information that may be learned, the analysis of several games of one team revealed that whenever it encountered a third down situation with the ball between its own fifteen and twenty yard lines on the right side of the field it ran its fullback on a draw play. If this information were available to its opponent, the linebackers could "key" on the fullback in this (and perhaps similar) situations to stop the play before it got started.

In addition to the standardized type of report described above, special reports may be requested according to the relevant needs of the moment. For example, if a coach wished to know by field position, down, and the distance for a first down, where the San Francisco Forty-Niners most frequently throw the screen pass, PROBE could provide the answer in a matter of seconds.

A critical aspect of the report-generation phase is the determination of the optimal manner of presenting the information to the user. One coach may prefer a graphical description of the data while another may prefer a detailed listing of successive events that occurred (such as all third down plays, in sequence, run from the left hash mark).

An ideal data bank would be one which includes all aspects of a game, both offensively and defensively, and which could be accumulated with appropriate weighting factors for data acquired in nonstandard situations. Weighting is necessary because a team may appear to have a strong passing attack when playing another team which has a weak pass defense or which has injured players. Another team may appear to be a blitzing team only because its opponent is particularly vulnerable to blitzing. The number of combinations of apparent strengths and weaknesses is large if a specific team is considered in the perspective of its play against other teams. All of these factors should be taken into consideration if accurate analysis is desired. Weighting would be an attempt to provide a standardization of all data accumulated on a specific team against all of its opponents.

Of course, the raw data without weighting would also be available for those coaches who desire it. It does provide a measure of reliable "gross"

information. Any football information system should have the capacity to
permit the retrieval of the tendencies of a team in a particular situation
and to calculate the probability of success or failure based upon previous
attempts in the same or similar situations. External factors such as weather, injuries, and other exogeneous playing conditions must also be considered. This type of information requires the continuous analysis of games
in order to maintain an information bank which is always current.

   As an alternative to using a "query" system which can be specialized
from team to team and from report to report, a team might rely on the same
"standard" set of reports for analyzing every opponent. If this alternative is chosen, a great deal of care must be exercised because the same
reports will be produced each week without regard to the nature of the opponent of the team being scouted or to the specific requirements of the
coach. For example, standard reports might be used to describe a team's
offensive tendencies without taking into account the defenses upon which
the offensive tendencies were shaped. In effect, little emphasis would be
placed on different circumstances, with misleading information as a possible result. This danger may be compounded if a coach is influenced by
the results of a specific report one week because of the favorable results
obtained from its use. The use of this report may well yield marginal or
even poor results in the future, however, because this future analysis would
be based on different criteria.

   Conclusion. The impact of the computer has been felt and the application of data processing to professional sports has developed quite rapidly
in recent years. The computer has been used in professional football in
processing scouting reports for individual players (for the pro draft of
college players) and for teams (strategy analysis as described above). The
present philosophy of planning in professional football is one of user-
preparation of statistics (that is, preparation by the coaches themselves)
and subjective decisions which are sometimes based on minimal information.
More detailed reports are often requested at more frequent time intervals.
It appears that there is in general a definite need for a more sophisticated
football data bank and professionally prepared reports for the use of coaches
in their decisions. Just as the manager relies on the information system to
collect and process data for decision making (his major function), the football coach would be free to coach without spending time collecting data and
preparing reports with a data bank which has been transcribed by an assistant. This would eliminate a substantial part of the timing problem in getting information to the players and would also enable the manager to make
his decisions on a timely basis.

A MODEL FOR EVALUATING PLAYER PERFORMANCE
IN PROFESSIONAL BASKETBALL

by

Bertram Price and Ambar G. Rao

Introduction. In a recent study, Noll and Okner [1973] found that the win-loss record of a professional basketball team has a significant impact on team profitability. The win-loss record is obviously dependent on team personnel and how they work together as a playing unit. This paper is an initial effort at examining team statistics to determine the characteristics that produce winning basketball with the goal of using these characteristics to evaluate trade, draft, or other personnel decisions.

Analysis of the data may be conducted at several levels. For instance one may look at team statistics, statistics of types of players (e.g. centers, guards, or forwards), or statistics of individual players. The data may be stratified by segment of season, by opponent, or by individual game. Undoubtedly all these forms of analysis would need to be performed before a reasonably complete characterization of winning basketball could be developed. In this paper we have focused on team statistics; the volume of such statistics is manageable and their analysis appears to be a natural starting point to an effort aimed at modeling player-acquisition decisions. The results of this study can be used as input for personnel decisions, perhaps in conjunction with the model developed by Price and Rao [1974].

Data Analysis. The discussion of team skills and their importance to winning in the NBA is an important matter for players, coaches, general managers, and millions of avid fans. Almost everyone with an interest in NBA basketball is able to give an analysis of the factors that contribute to winning; often that analysis is based on years of careful observation and intense study. The solutions range. It is the offense-minded big man--Mikan, Jabbar, McAdoo, Chamberlain (of old); or the defense-minded big man--Chamberlain (of new). It could be the super-forward, Arizin, Petit, Barry, Baylor; or a playmaking guard, Cousey. It may be the complete player, West, or a synergistic collection of skills and partial skills such as the more recent versions of the New York Knicks. Some analysts may seek the solution from data found in the record book. A more general approach using data is to ask what characterization of winning can be developed by an objective analysis of statistics and performance measures of past seasons? This is our approach and we hope that our personal biases concerning big men, playmakers, and synergism have been minimized in the analysis.

The analysis is carried out on an aggregate level, using team statistics. The variables used and their abbreviated names appear in Table I. The first eight variables are treated as the basic entities that determine success. Success is defined either as win-loss percentage or participation in the postseason playoff series. We shall attempt to establish some transformations of the eight basic variables that most accurately predict success.

Data and Variables. Official NBA data going back to the 1953 season was supplied for the variables listed in Table I by the offices of the National Basketball Association. The first eight variables are taken as the basic set of independent variables in the sense that they are the basic

116

Table I

Variable	Abbreviation
Field Goals Attempted per Game	FGA
Field Goals Made per Game	FGM
Free Throws Attempted per Game	FTA
Free Throws Made per Game	FTM
Rebounds per Game	REB
Assists per Game	AST
Personal Fouls Committed per Game	PF
Disqualifications per Game	DQ
Win-Loss Percentage	PCT
Playoff Indicator	PLF

1 - if in playoff, 0 - if not in playoff

Table II

Variable	Abbreviation
Field Goal Percentage	FGP
Free Throw Percentage	FTP
Field Goals Made per Assist	FGMA
Field Goals Attempted per Rebound	FGAR
Free Throws Attempted per Personal Foul Committed	FTAPF
Field Goal-Free Throw Interaction	FGP x FTP
Rebound-Assist Interaction	REB x AST

determinants of success. A set of transformed variables that were thought to provide some additional insight into the characterization of success were also developed. These variables are listed in Table II. The rationale for some of these variables such as FGP and FTP is obvious. Others require a word of explanation. FGP x FTP is an attempt to identify a possible non-linearity in the contributions of field goal accuracy and free throw accuracy. Similarly REB x AST is a nonlinear term that is related to the value of rebounds and assists. FGMA represents an indirect attempt to identify the value of assists. If a hypothesis were in order at this point, it would state that low values of FGMA were associated with success. That is, field

goals that result from assists are high-percentage shots which also suggests
a savings of individual energy in point production that may be transferred
to other aspects of the game. FGAR is an attempt to measure the efficient
use of rebounds. A rebound taken on defense represents possession of the
ball and a change to offense. An effective use of the rebound would be to
transform it into at least one attempted field goal. The same argument can
be applied to offensive rebounds. Therefore FGAR should have a positive
correlation with success. FTAPF measures the balance between the potential
for yielding points due to personal fouls committed against opponents and
the potential for scoring points due to personal fouls committed by opponents. Each variable was tabulated by year for each team. A data observation was defined as a vector of 17 values (2 success measures, 8 basic
variables, 7 transformed variables) for a particular team in a given year.
Pooling the data for 1953-1973 results in a data set consisting of 227 such
vectors.

Because the NBA brand of basketball has changed both in rules and style
over the past 21 seasons, it is clear that some form of data editing is required prior to the statistical analysis. Some rule changes are difficult
to evaluate. For example there have been many changes in rules concerning
the shooting of free throws. The most recent change awards possession of
the ball rather than awarding a free throw for certain fouls. It follows
that a typical foul committed in 1959 may not have the same implication
for a successful season as a typical foul committed in 1973. The impact
of other rule changes such as expanding the three-second lane or introducing a goal-tending violation may be equally difficult to assess. We assume
that a rule change or a change in the style of play is inconsequential to
our objective unless the change is directly reflected in the variables under
analysis.

The most obvious change in the data is length of season. The number
of games played has increased from 72 to 82. Therefore, each variable was
tabulated on a per-game basis. To see what other adjustments were necessary, the yearly league averages were tabulated for each of the eight basic
variables and are displayed in Figures 1 and 2. We see that FGA is unstable
especially in the early years of the data. Undoubtedly some of the in-
stability is due to the 24-second rule which was first used in the 1954-55
season. Both FGA and FGM appear to stabilize around 1961 or 1962, although
there also appears to be a trend in very recent seasons of lower FGA, constant FGM, and thus improved FGP. AST is also unstable in the early years
of the data set. It appears to become stable after 1964. FTA has been
affected in the past two seasons due to the rule change that awards possession instead of a free throw.

Various adjustments directed at eliminating the season-to-season variation of the variables were considered. The initial exploratory analyses
were performed on data that had been standardized using yearly means and
standard deviations. Standardized data was not used for estimating model
parameters because of the relative difficulty of interpreting models that
are stated in adjusted variables. Rather, it was decided to foresake the
full 21 years of data and use subsets of raw data that showed stability.
Data prior to 1968 was discarded. The remaining data were analyzed as two
separate subsets formed by the years 1968 to 1971 and 1972 to 1973.

Statistical Analysis. Four multivariate statistical analysis technique were applied to the data. Factor analysis was used with the stan-

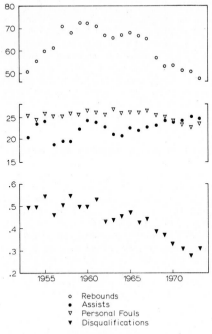

o   Rebounds
•   Assists
▽   Personal Fouls
▼   Disqualifications

Figure 1

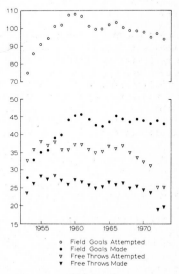

o   Field Goals Attempted
•   Field Goals Made
▽   Free Throws Attempted
▼   Free Throws Made

Figure 2

dardized data in an exploratory analysis directed at identifying composite
variables that could better summarize the variation in the independent
variables. Automatic Interaction Detection [Sonquist, et al, 1971] was
also used in an exploratory mode to try to identify interactions in the
independent variables that could effectively explain variation in PCT.
Discriminant analysis was used to identify the variables that were most
important in distinguishing between teams that make it to the postseason
playoffs and those that do not. Regression analysis was used to establish
the relationship between PCT and the important independent variables as
well as to estimate the relative contribution of each variable. The results of all four analyses were fairly consistent in identifying the more
important variables.

Factor analysis results were obtained for the principal-components
model, the orthogonal-factor model, and the oblique-factor model. In each
case the factors extracted were identified with the basic variables. As
an example see Table III. Clearly the loadings indicate that there is not
much difference between the factors and the basic variables. Automatic
Interaction Detection showed FTP, REB, and AST to be important determinants for explaining the variations in PCT. The result suggests what is
well known by most successful coaches, namely that games are won and lost
in the less glorious pursuits of free throws, rebounds, and assists. The
discriminant analysis held no surprises. FGP, FTP, REB, and PF were the
important variables for discriminating between participating and not participating in the playoffs. The discriminant functions were generally
able to predict playoff participation correctly for about 75% of the observations. None of the analyses discussed above clearly identified any
new or special composite variables. As a result, regression analysis was
performed using the 15 independent variables as they were originally defined. It is gratifying that the significant explanatory variables are
almost all simple and easy to interpret. This makes the results of the
statistical analysis easier to use in the decision-making context discussed
below.

PCT was regressed against various combinations of the 15 independent
variables in search of a subset of effective predictors. The results of
the analysis identified PF, REB × AST, and FGP × FTP as highly significant.
The presence of the two nonlinear (multiplicative) variables suggests the
possibility of a fully multiplicative model. The idea of a multiplicative
model was pursued because of its relative simplicity. The effect of a
variable such as FGP × FTP would be present but the decision maker would
only be required to interpret and manipulate FGP and FTP separately. The
multiplicative model takes the form

$$PCT = a_0 FGP^{a_1} FTP^{a_2} PF^{a_3} REB^{a_4} AST^{a_5}$$

The parameters were estimated for data from 1968 through 1971 and data from
1972 through 1973. The resulting coefficients appear in Table IV.

The multiplicative model is in the form of the Cobb-Douglas type production function. PCT may be viewed as output with FGP, FTP, PF, REB, and
AST as the factors of production. The regression coefficients may be interpreted as factor elasticities; for example, for the 1972-73 regression a
1% change in FGP results in a 2.05% change in PCT.

Table III

Orthogonal Factor Analysis Factor Loadings

Variables	\multicolumn{8}{c}{Factors}							
	1	2	3	4	5	6	7	8
FGA	.14	-.17	-.06	-.15	.07	.95	.03	.06
FGM	-.11	.04	.01	.68	-.07	.68	-.01	.01
FTA	-.08	.96	-.16	.09	.02	-.07	-.09	.06
FTM	-.08	.88	-.08	.07	.08	-.05	.41	.06
REB	-.03	.06	-.77	.06	-.08	.61	.02	.01
AST	-.92	.04	-.04	.28	-.07	.23	.00	-.01
PF	.09	.04	.12	-.07	.94	.00	.08	.29
DQ	.09	.05	.03	-.04	.30	.06	-.04	.94
FGP	-.29	.19	.03	.91	-.09	-.15	.02	-.06
FTP	.02	.08	.05	.01	.08	.02	.99	-.04
FGMA	.97	-.06	.06	.11	.03	.13	.02	.06
FGAR	.17	-.26	.90	-.19	.15	.09	.02	.04
FTAPF	-.12	.70	-.19	.11	-.59	-.06	-.11	-.15
FACTOR ID	AST	FT	REB	FGP	PF	FGA	FTP	DQ

The coefficients are not constant; the later data show: (i) a slight decrease for FGP and a corresponding increase for FTP; (ii) a sizable increase in the negative contribution of PF and (iii) dramatic changes in the marginal contributions of REB and AST. The differences have some plausible explanations. Both the style of play and the rules have undergone change in the 1968-1971 time frame. In general, defensive aspects of the game have been emphasized and the concept of team play has become more important. AST and FGP have been increasing and REB has been decreasing. There has been a major rule change concerning fouls that replaces a free throw with possession of the ball. The increase in the magnitude of the coefficient of PF may be directly related to the rule change. As a side effect, since there are fewer free throws attempted, their marginal contribution is magnified. Also, possession of the ball due to fouls may be competing in importance with possession due to rebounds. Since possession is worth about one point, and a turnover thus about two points, the increase in both FGP and AST may serve to explain the reduction in the marginal contribution of REB. In general, the contributions of the variables to success will necessarily change as the game changes. The statistical analysis will have to be supplemented with some judgments if it is to be useful for making trading and drafting decisions.

Table IV

Regression Results for Multiplicative Model

	1968-1971		1972-1973	
Variable	Regression Coefficient	t	Regression Coefficient	t
Constant	-4.21		820.58	
FGP	3.21	5.38	2.05	1.98
FTP	2.18	2.32	3.77	3.02
PF	- .98	-2.14	-2.74	-4.66
REB	2.47	4.91	.54*	1.58
AST	-	not significant	.54*	1.58
	$R^2 = .50$		$R^2 = .57$	

*Empirical analysis suggested that REB and AST should have equal
weight. Regressions on the 1972-73 data that treated REB and
AST as separate variables resulted in both regression coefficients
with t values considerably less than 1. When the regression is
constrained to impose equal weight, the resulting coefficient
is at least marginally significant.

The findings of the statistical analyses and the type of judgmental
information that is required can be used in a decision calculus type model
[Little, 1970] to help make decisions about trading and drafting. The
model can help identify team weaknesses or personnel gaps and be used to
analyze alternative remedies. The model would incorporate team characteristics, as well as personnel costs and how they affect performance. For
a discussion of this approach see Price and Rao [1974].

    Conclusions. Aggregate data on team statistics in the National Bas-
ketball Association were analyzed to determine the most important factors
contributing to a team's win-loss record. A multiplicative model was estimated using regression analysis. Using this information, and other factors
which we believe to be important to the decision (but not yet captured
through the data analysis), it is possible to develop a model for the personnel decisions. It is suggested that some of the missing parameters may
be estimated through a decision calculus approach. That work is continued
in Price and Rao [1974].

    The next steps toward the goal of an operational model are: (1) more
detailed analysis of basketball statistics, possibly at the player-by-game
level, to establish the forms of many of the functions used in the model;
(2) fitting the model to actual data, using whatever judgmental inputs are
necessary; and (3) using the model for predictive purposes, and adjusting
coefficients and/or functional form if so indicated.

PLAYOFF STRUCTURES IN THE NATIONAL HOCKEY LEAGUE

by

James P. Monahan and Paul D. Berger

Introduction. Every sport is played within the boundaries of certain well-defined rules and procedures. Although infrequently acknowledged, the structural dimensions of a contest often play a large role in shaping the eventual winner [David, 1959; Glenn, 1960; Searls, 1963]. One of the clearest examples of this phenomenon occurred a few seasons ago in the sport of professional ice hockey.

In 1971 many National Hockey League (NHL) fans were surprised and out-raged when the Boston Bruins, reputedly the best hockey team in many years, were dramatically upset by the Montreal Canadiens in the first round of the Stanley Cup Playoffs. This unexpectedly early elimination caused consider-able concern about the "overall fairness" of the league's playoff structure. The Bruins set 35 new major-league hockey scoring records, outdistanced all other teams by posting a record of 57 games won, 14 lost, and 7 tied, and yet were denied the championship because of losing a seven-game series, 4-3.

Partly in response to this outcry, the NHL has since introduced a new playoff structure. In this paper, we take a comparative look at the old playoff structure (through 1970-71) and the new playoff structure, beginning 1971-1972 and still in effect as of this writing. We also propose and ex-amine some other playoff structures which perform better than the two above structures, relative to likely criteria of the average hockey fan.

The NHL. The NHL in 1971-1972 was comprised of 14 teams split into two 7-team divisions, East and West. In 1972-73 one additional team was added to each division. During the regular season, each team plays 78 games against the other teams. The winner of each game receives 2 points, the loser 0 points; if the games end in a tie, each team receives 1 point. (Thus, the Bruins, with a (W-L-T) record of (57-14-7) in 1971 earned 57(2) + 7(1) = 121 points). At the end of the season, the 4 teams with the highest point totals within each division enter Stanley Cup playoff com-petition.

These 8 teams are systematically paired off in quarter-finals compe-titions, the winners systematically paired off in semifinals, and the two teams which emerge enter the finals. The winner of a pairing off is the team which first wins 4 games. (There are no ties in playoff competition. In the event of a tie, "sudden death" overtime is invoked in which the first goal wins). How the 8 teams are systematically paired off is re-ferred to as a "Playoff Structure".

Playoff Structure. In the first round of the old playoff structure, only intradivisional matchings occurred. Within each division, the teams ranked 1 and 3 were paired and the teams ranked 2 and 4 were paired. In the semifinal competition, two interdivisional series were played, the win-ner of a (1-3) pairing in one division playing the winner of a (2-4) pair-ing in the other division. The winners of these series then entered the Cup finals.

Under the rules of the new playoff structure, once again no inter-

Figure 1
Old and New Playoff Structures
(E = East, W = West, and the Numbers Refer to Ranking Within Division)

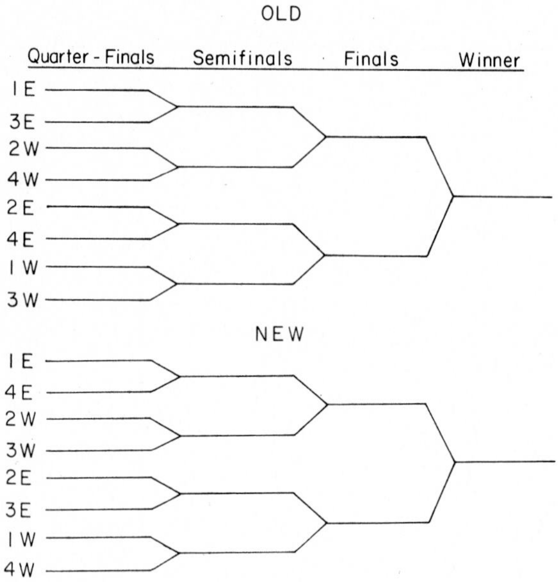

divisional matchings occur in the quarter-finals, but now within each division, teams 1 and 4 are paired and teams 2 and 3 are paired. In the semifinals, the winner of a (1-4) pairing in one division plays the winner of a (2-3) pairing in the other. The ensuing winners meet in the Cup finals.

We present a schematic of the two structures in Figure 1.

Criteria. We suggest three criteria against which to measure the suitability of a playoff structure. The first criterion is to maximize the probability that the highest ranked team of the 8 (i.e., the team with the most points) wins the Stanley Cup. The second criterion is to maximize the expected number of points of the cup winner; i.e.,

$$\max W = \sum_{i=1}^{8} P(\text{team } i \text{ wins the cup}) \cdot (\text{Points of team } i). \qquad (1)$$

This second criterion allows for cases where two or three teams are far above the others, and it might seem unfair to use criterion 1, judging a structure according to how it treats a single team whose point total could be less than 1% above the point total of the second-ranked team. In our opinion the second criterion is reasonable in that it tends to "spread" the advantage according to the team's relative "deservedness". A third criterion is that the two best teams should meet in the finals.

Table I.

Win Probabilities for Playoff Teams in 1971.
First row, p(i wins one game from j).
Second Row: p(i wins best of seven from j)

Team i \ Team j	N.Y.	Montreal	Toronto	Chicago	St Louis	Phil.	Minn.
Boston E1, 121 pts.	.5260 .5569	.5550 .6190	.5960 .7025	.5307 .5670	.5817 .6741	.6237 .7546	.6269 .7605
New York E2, 109 pts.		.5291 .5635	.5707 .6516	.5046 .5101	.5561 .6212	.5989 .7081	.6022 .7145
Montreal E3, 97 pts.			.5419 .5910	.4755 .4465	.5272 .5592	.5706 .6514	.5739 .6583
Toronto E4, 82 pts.				.4339 .3578	.4852 .4677	.5290 .5632	.5324 .5707
Chicago W1, 107 pts.					.5515 .6116	.5944 .6993	.5978 .7058
St. Louis W2, 87 pts.						.5437 .5950	.5471 .6023
Philadelphia W3, 73 pts.							.5034 .5075
Minnesota W4, 72 pts.							

Table II

Team	Points	P(winning Cup) Old Structure	P(winning Cup) New Structure
E1, Boston	121	.2641	.2944
E2, New York	109	.2005	.1761
E3, Montreal	97	.1106	.1081
E4, Toronto	82	.0572	.0581
W1, Chicago	107	.2014	.1967
W2, St. Louis	87	.0951	.0960
W3, Philadelphia	73	.0343	.0400
W4, Minnesota	72	.0368	.0306
		1.0	1.0

We proceed to evaluate the old and new playoff structures as well as other structures which perform better than either of these two playoff structures, according to these criteria. In all cases, we use 1971 data as our base, and our probabilities of winning the Cup refer to a fictional reissue of the 1971 Stanley Cup.

A Probabilistic Model. There are numerous ways to estimate the probability that team i will win a single game from team j. One way is to base the estimate on the results of the regular-season games between the two combatants. Another way is to base the estimate on the results of the full regular-season schedule. A third way would be to develop a complete probabilistic model which considers a distribution of time between goals scored, a distribution of time between goals scored upon, assumptions about independence or lack thereof between offense and defense, adjustments for situations in which the goalie is removed for an extra skater, etc. Other ways might be based on point totals, goals scored, Bayesian subjective assessments, etc.

The first estimation procedure suffers from the small sample size-- actual estimation would be based on only six games. The third estimation procedure probably would yield the most accurate estimate, but was rejected because of its operational complexity. We chose the second method. Thus we estimate the probability that team i will win a single game from team j in Stanley Cup competition by

$$P(i \text{ over } j) = \frac{\text{Total points of team } i}{\text{Total points of team } i + \text{total points of team } j} .$$

While this formula underestimates the correct probability if team i has won nearly all its games, it is probably close in most realistic cases, and leads to the correct ranking of various playoff structures under criteria 1 and 2.

We assume a Bernoulli process, similar to Mosteller's analysis of World Series competition [1952]. Under our Bernoulli model, we are assuming that performance in any game at any stage of the playoffs is independent of past performance in these playoffs, but is determined by results throughout the regular season. Table I gives the probability that team i beats team j, in a single game and in a "best four out of seven". The latter is derived from the formula:

$$P(i \text{ wins 4 games before } j) = \sum_{k=0}^{3} [P(i \text{ over } j)]^4 \cdot [1-P(i \text{ over } j)]^k \cdot \binom{3+k}{k}$$

[Mosteller, 1952].

Results. The probability that any given team wins the Cup can now be found by using the numbers in the second rows of Table I, the appropriate playoff structure (Figure 1), and some rules of elementary probability theory. Table II gives the probability that each team wins the Cup given the old and new playoff structures. Table II also repeats, for convenience, the yearly points of each team.

Evaluating the two structures according to criterion 1, observe the relatively higher probability under the new structure of E1 (the team with

the most points) winning the Cup. Thus, according to this criterion, the
new structure is preferable. Translated into the language of a sports fan,
our analysis says that if the new structure had been available in 1971, the
Bruins would have had a better chance of winning the Cup than they actually
were afforded under the old system.

In order to evaluate the old system according to criterion 2, we must
look at the inner product of columns 1 and 2 in Table II, which is 104.2062
To evaluate the new system according to criterion 2, we look at the inner
product of columns 1 and 3 in Table II. This gives 104.5893. Comparing
the magnitude of the two figures, we once again find that the new playoff
system is slightly preferred.

Some Other Structures. One proposed scheme is shown in Column (a) in
Figure 2. The teams are ranked from 1 to 8 by point standings without regard to division. This is the seeding system most frequently used in
major knockout tournaments, team-of-four bridge tournaments, tennis matches,
collegiate hockey, etc.

Figure 2

Alternative Playoff Structures: a, b, c

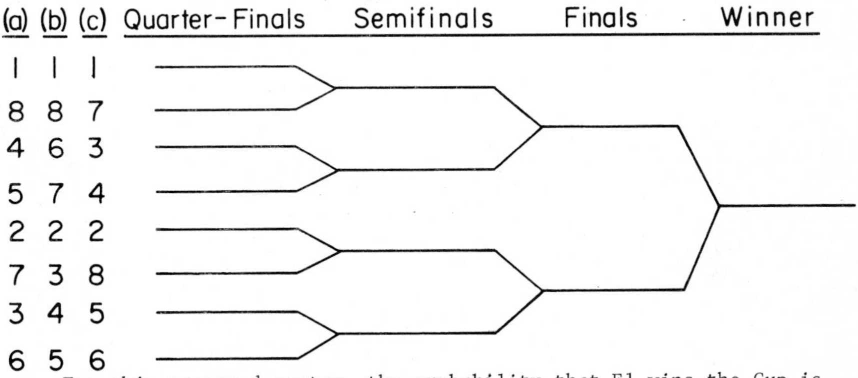

For this proposed system, the probability that E1 wins the Cup is
.2945, which is slightly higher than under either of the previous two playoff systems. In addition, the value of criterion 2, W, is 105.3305. According to the second criterion this proposed playoff scheme is preferable
to either of the two systems.

Column (b) of Figure 2 presents a structure which maximizes the probability that the best team (team 1) wins the cup. We have not been able
to prove formally that this optimal structure is unique. However, extensive computer simulations have found it to be unique in all trial cases.
In these simulations we used as input sets "consistent probability matrices",
which we defined as an 8 x 8 table in which each off-diagonal cell P_{ij},
$i \neq j$, is the probability that team i beats team j in a seven game series;
P_{ij} is strictly monotonic increasing as j increases for fixed i; P_{ij} is
strictly monotonic decreasing as i increases for fixed j; and $P_{ij} = 1 - P_{ji}$.

Column (c) of Figure 2 illustrates a playoff structure which fre-

quently maximizes the probability that the two best teams meet in the finals
(a promoter's dream). This structure is not optimal for all consistent
probability matrices; however, in the vast majority of simulation trials it
was optimal. One of the simplest descriptions of a consistent probability
matrix is when $P_{ij} = j/(i+j)$; for this matrix it can be shown that the
structure in Column (c) is indeed optimal. (We are not suggesting that
$P_{ij} = j/(i+j)$ is necessarily representative of the real world). To illustrate that the structure in Column (c) is not always optimal by this criterion, consider the case where the P_{1j}'s are all approximately equal, but
the P_{2j}'s vary considerably. Then it is clear that team 2 should have
teams 6, 7, and 8 in its bracket.

   Aftermath. We discussed the question of an "equitable" playoff structure with the owner of one of the NHL teams. He acknowledged that not
enough thought has been given to the objectives (criteria) of a playoff
structure. He did, however, point out to us that while the criteria we
have considered are all reasonable, we have neglected an important factor--
namely money. In some of our proposed structures a lower-ranked team has
a higher chance of reaching the semifinals and finals than some higher-
ranked teams. Since the players are paid according to how far they sur-
vive in the playoffs (and the owners according to total attendance), some
of our proposed structures would not be deemed equitable by them.

                                    Summary

   Although infrequently analyzed, the structural dimensions of a contest often play a large role in shaping the eventual winner. In this
paper, we take a comparative look at the new and the old playoff structures of the National Hockey League's Stanley Cup Competition. Our concern is with the "overall fairness" of each structure and we therefore
propose and examine some other playoff structures which perform better
than the two above structures, relative to likely criteria of the average
hockey fan.

CRICKET AND STATISTICS

by

Richard Pollard

Historically, cricket holds a unique position in the development of
the scientific study of sports and games. One of the first books on statistical method [Elderton, 1909] made extensive use of cricket to illustrate elementary theory. Some years later G. H. Wood published a series
of articles in The Cricketer, which culminated in the reading of a paper
to the Royal Statistical Society [Wood, 1945]. At the same meeting, Sir
William Elderton also read a paper on cricket [Elderton, 1945], so that he
and Wood must share the distinction of producing the first full quantitative
paper on sport. Wood died a few months later and with him appears to have
vanished all interest in the application of statistics to cricket. Not a
single paper has appeared since, despite a wealth of factual information
available on England's national sport. The work of Elderton and Wood can
be summarized under three main headings: distribution theory, consistency,
and skew correlations.

Distribution Theory. No satisfactory distribution has been found to
describe the frequency of scores of individual batsmen at cricket. Elderton [1927] first used the scores of the Yorkshire batsman Tunnicliffe to
illustrate a Type X Pearson curve, which is exponential. In a later paper
however [Elderton, 1945], the same set of scores is presented as a geometric
distribution; this is not surprising, since the geometric distribution is
the discrete analog of the negative exponential, and for a large mean the
two approximate one another very closely. Reep et al [1971] showed that
some series of scores produce a close fit to the negative binomial distribution. This distribution has the geometric distribution as a special case.
In order to fit the negative binomial distribution, 2 parameters, k and p,
are estimated from the observed data. If k is assumed equal to 1, then the
geometric distribution is obtained, the fitting of which is done by estimating only p. For Elderton's data, a comparison of these two fitting
procedures is shown in Table 1. The geometric distribution gives slightly
closer fits, but closer inspection of the frequency distributions shows
that both fits suffer from the same drawback: the variance of the observed scores is higher than that of either theoretical distribution. This
confirms a common belief in cricket: there is a high likelihood of a batsman being dismissed early in his innings before he is settled, but once
settled, adding to an already sizeable score becomes easier. It seems,
therefore, that a more elaborate model needs to be developed to describe
the distribution of batsmen's scores, and to discover the exact nature of
the random process which is at work.

Consistency. Two different meanings have been attached to the term
consistency as applied to batsmen's scores at cricket. Elderton [1909]
used the C.V. (coefficient of variation = 100 x standard deviation/mean)
to give a measure of consistency. The lower his C.V., the more consistent
the batsman. Wood [1945] argued that since the "geometric progression"
described batsmen's scores, the closer his scores were to this distribution, the more consistent the batsman. The coefficient of variation for
a geometric distribution is almost exactly 100 (actually $100(x - 1)^{\frac{1}{2}}x^{-\frac{1}{2}}$
where x is the mean; thus C.V. > 98 for batsmen who average 25 or more runs).

Hence the nearer his value to 100, the more consistent the batsman.   Thus
a batsman who always scores exactly 20 is totally consistent for Elderton,
but highly inconsistent for Wood.   Since the geometric distribution is only
an approximation to the real state of affairs, Wood's definition needs re-
vision, although his own concept of 'consistency' remains a valid one.

   Skew Correlations.   In cricket, a team's innings is started by two
batsmen who remain together until one is dismissed.   Since the individual
innings of these two batsmen begin under identical conditions, Elderton
[1945] suggested that there might be some correlation between their scores.
Analysis of the scores of several pairs of batsmen, however, failed to
demonstrate any significant correlation.   Elderton noted that since the
distribution of a single batsman's scores was markedly skew (i.e., roughly
geometric) and certainly not normal, any correlation would be hard to dis-
cern.   After developing several other skew distributions, Elderton indicated
how they might be applied to cricket scores.

Table I

Goodness of Fit Tests of the Distribution of Batsmen's Scores

Batsman	Negative Binomial		Geometric		Average number of runs
	x^2	$p(> x^2)$	x^2	$p(> x^2)$	
Hayward	2.73	.74	1.44	.96	54.6
Jessop	6.62	.08	6.14	.19	29.8
Tunnicliffe	0.71	.87	0.31	.99	30.2
Warner	5.58	.13	5.99	.20	41.0

THE MOST IMPORTANT POINTS IN TENNIS

by

Carl Morris

Introduction. What are the most important points in tennis? Professional players and coaches have given many answers to this, among them: the first point (it's important to get off to a good start); 15-all (it's pivotal); the deuce and advantage points. Most frequently 15-30 and 30-15 are singled out; for example recently by coach Bob Harman [1976]; long ago by Bill Tilden; and in between by Pancho Gonzales. Gonzales emphasized the importance of the 15-30 point because if the server wins, the score will be 30-30 and he'll probably go on to win the game; but if he loses, the score will be 15-40 and he's likely to lose the game.

This formulation of the concept of "important", comparing the probability of winning the game with respect to winning and losing the given point, is evident in almost every argument on the subject, and is formalized here. The $\underline{importance}$ of a point for winning a game is defined to be the difference between two conditional probabilities: the probability that the server wins the game given that he wins the point, minus the probability that he wins the game given that he loses the point. In mathematical terms: $I_{sr} = P_{s+1,r} - P_{s,r+1}$ where I_{sr} is the importance of the point when the server has score s and the receiver score r, and P_{sr} is the probability that the server will win a game in which the score is s to r.

Since the receiver's probabilities are the complements of the server's, the same definition, if applied to the receiver, would yield the same difference between his two probabilities, and hence the same numerical importance. Hence: Every point is equally important to both players. This statement is true of every "two-person, zero-sum game", a property which characterizes most athletic contests. This concept is emphasized here because of the popular misconception that points are more important to the player or team which is behind.

It is remarkable that the Tilden-Gonzales-Harman conjecture, when translated into the language of mathematics, is easily shown to be incorrect. Counting the tennis scores of 15, 30, and 40 as 1, 2, and 3 points respectively, the above definition of importance yields, for 15-30, $I_{12} = P_{22} - P_{13}$, and for 30-40, $I_{23} = P_{33} - P_{24}$. Clearly $P_{22} = P_{33}$, since in either case two points are required to win the game; $P_{24} = 0$, since the server has zero probability of winning the game, having already lost it; and $P_{13} > 0$. It follows that $I_{23} > I_{12}$. Thus 30-40 must be a more important point than 15-30.

It is advantageous for a tennis player to know which points are more important, since a player can win with well under half the points, provided he dominates the points that count most. Consequently, on the important points, astute players concentrate and play hardest, hit their best shots, and exploit their opponent's weaknesses. On relatively unimportant points they are more likely to save energy or attempt to force their opponents to expend energy, to vary tactics and pace to keep the opponents off balance, and to use this opportunity to disguise their own strengths and tactics.

Probabilistic Model. We assume that the server wins each point with probability p, $0 \leq p \leq 1$, and loses the point with probability q = 1-p. Successive points are independent, i.e. p does not depend on the outcome of earlier points. This last assumption, which ignores the effects of fatigue, has little effect on the conclusions reached below. We frequently assume that $0.5 \leq p \leq 1$; this is true in all good tennis, and in any case causes no loss in generality since results for $p < 0.5$ are obtained by interchanging the server and the receiver. For equally matched professionals, p = 0.6 is probably typical. It varies with individual players and type of court surface, and increases for high-level doubles play. The weakest part of the model is the independence assumption, but this assumption, which has been used by others [Carter and Crews, 1974; Kemeny and Snell, 1960] leads to a useful theory of importance.

Preview of Some Results. The main results are contained in Tables I and II which present points, ranked by their importances, together with the numerical values of their importances, their efficiencies, and other infor- mation, for p = .5 and .6 respectively. The importance of points depends heavily on the probability p of holding serve, as does the variation of importances. For $p > .5$, the most and least important points are 2-3 and 3-0 respectively. The ratio of their importances is 4 for p = .5, 14.06 for p = .6 and 144 for p = .75. The second most important point is 2-2 (of course $I_{22} = I_{33}$) if $p < .62$ and (surprisingly) 1-3 if $p > .62$. If p = .5 exactly, then 2-3, 2-2, 3-3, and 3-2 are all equally important. The score 1-2 never ranks higher than third for any $0 < p < 1$, while 2-1 is rela- tively unimportant for $p > .5$, and become less important as p increases.

The purpose of this paper is to explain these tables and the "Lorenz curve", Figure 1, which partially summarizes them. A striking result of Figure 1 is that more than half of the increase in probability of winning a game due to an increase in effort can be achieved by confining the extra effort to less than half the points--those points which are most important. Identification of these points and their exact numerical values is revealed in Tables I and II; if the server has a distinctly better than even chance of winning each point, then the differences are dramatic, as we shall see.

Finally, some games are more important than others for winning a set, and some sets (e.g., the third one in a three-set match) are more important for winning a match than others. The idea developed for the importance of points to winning games may be extended in a consistent manner, but with necessary new wrinkles, to compute the importance of games to winning sets, and of sets to winning matches, and this is sketched at the end of this article.

Some notation, but little mathematical theory, is presented in the text. Certain mathematical results, which permit construction of tables and figures for other values of p, and which provide the foundation upon which the assertions of the text follow, are given in the appendix without proof.

Numerical Computation of Importance of Points for Games. Let p be the probability that the server wins a point, and be unchanged from point to point throughout a game, with q = 1 - p. If the server has s points and the receiver r points, the probability, P_{sr}, that the server will win the game can be computed from

$$P_{sr} = pP_{s+1,r} + qP_{s,r+1}$$

Table I

Importance and Other Data for Points Ordered by Their Importance
p = 0.5 (L = 6.750, M = 2.500, $\overline{I}$ = .3704)

1	2	3	4	5	6	7	8	9	10	11
s	r	P	N	f	F	I	T	G	E	e
2	3	.2500	.625	.0926	.0926	.5000	.3125	.1250	1.350	1.350
2	2	.5000	.375	.0556	.1481	.5000	.1875	.2000	1.350	1.350
3	3	.5000	.625	.0926	.2407	.5000	.3125	.3250	1.350	1.350
3	2	.7500	.625	.0926	.3333	.5000	.3125	.4500	1.350	1.350
1	2	.3125	.375	.0556	.3889	.3750	.1406	.5063	1.302	1.013
1	1	.5000	.500	.0741	.4630	.3750	.1875	.5813	1.256	1.013
2	1	.6875	.375	.0556	.5185	.3750	.1406	.6375	1.229	1.012
0	1	.3438	.500	.0741	.5926	.3125	.1563	.7000	1.181	.844
0	0	.5000	1.000	.1481	.7407	.3125	.3125	.8250	1.114	.844
1	0	.6563	.500	.0741	.8148	.3125	.1562	.8875	1.089	.844
1	3	.1250	.250	.0370	.8519	.2500	.0625	.9125	1.071	.675
0	2	.1875	.250	.0370	.8889	.2500	.0625	.9375	1.055	.675
2	0	.8125	.250	.0370	.9259	.2500	.0625	.9625	1.040	.675
3	1	.8750	.250	.0370	.9630	.2500	.0625	.9875	1.025	.675
0	3	.0625	.125	.0185	.9815	.1250	.0156	.9938	1.013	.338
3	0	.9375	.125	.0185	1.0000	.1250	.0156	1.0000	1.000	.338

LEGEND: (s,r) = server's and receiver's scores. P_{sr} = probability of winning from (s,r). N_{sr} = expected number of times point played at (s,r). f_{sr} and F_{sr} = fraction and cumulative fraction of all points played at (s,r). I_{sr} = importance of point (s,r) to winning the game. T_{sr} = time-weighted importance. G_{sr} = cumulative gain. E_{sr} = cumulative efficiency. e_{sr} = efficiency of effort expended at (s,r). Full explanation in text.

Table II

Importance and Other Data for Points Ordered by Their Importance
p = 0.6 (L = 6.484, M = 2.086, $\overline{I}$ = .3217)

1	2	3	4	5	6	7	8	9	10	11
s	r	P	N	f	F	I	T	G	E	e
2	3	.4154	.443	.0683	.0683	.6923	.3067	.1471	2.152	2.152
2	2	.6923	.346	.0533	.1216	.4615	.1595	.2235	1.838	1.435
3	3	.6923	.532	.0820	.2036	.4615	.2454	.3412	1.675	1.435
1	2	.5151	.288	.0444	.2480	.4431	.1276	.4024	1.622	1.377
1	3	.2492	.154	.0237	.2717	.4154	.0638	.4329	1.593	1.291
0	2	.3689	.160	.0247	.2964	.3655	.0585	.4610	1.555	1.136
0	1	.5762	.400	.0617	.3581	.3456	.1382	.5273	1.472	1.074
1	1	.7145	.480	.0740	.4321	.3323	.1595	.6037	1.397	1.033
3	2	.8769	.665	.1025	.5346	.3077	.2045	.7018	1.313	.956
0	0	.7357	1.000	.1542	.6888	.2658	.2658	.8292	1.204	.826
2	1	.8474	.432	.0666	.7555	.2585	.1117	.8827	1.168	.803
0	3	.1495	.064	.0099	.7653	.2492	.0160	.8904	1.163	.775
1	0	.8421	.600	.0925	.8579	.2127	.1276	.9516	1.109	.661
2	0	.9271	.360	.0555	.9134	.1329	.0479	.9745	1.067	.413
3	1	.9508	.346	.0533	.9667	.1231	.0425	.9949	1.029	.383
3	0	.9803	.216	.0333	1.0000	.0492	.0106	1.0000	1.000	.153

with boundary conditions $P_{04} = P_{14} = P_{24} = 0$, $P_{40} = P_{41} = P_{42} = 1$, and $P_{22} = P_{33} = \ldots = p^2/(p^2 + q^2)$. Note that the probability of winning from deuce is the probability that the server wins two consecutive points given that someone wins two consecutive points. These formulas may be evaluated recursively by computer, or by the closed-form expressions in the appendix. The results are shown in the first three columns of Tables I and II.

Column 4 of the tables shows N_{sr}, the expected number of times per game that the point (s,r) is played. $N_{00} = 1$, since every game starts with the server and receiver having no points. In most cases N_{sr} is the probability that the score ever reaches (s,r), because most scores may be reached only once. But the scores 2-3, 3-3, and 3-2, which represent ad-out, deuce, and ad-in, are exceptions. Formulas for N_{sr} appear in the appendix.

The sum of the values N_{sr} is denoted by L, and is the expected number of points in a game. The ratio $f_{sr} = N_{sr}/L$ (column 5) is the expected fraction of all points that will be served with the score being (s,r). The values F_{sr} (column 6) are the cumulative values of f_{sr}. Since the points are ordered by importance, F_{sr} represents the total expected portion of points that are at least as important as (s,r). These values are used to construct the horizontal axis of the Lorenz curves in Figure 1.

Column 7 contains values of the importance, defined as $I_{sr} = P_{s+1,r} - P_{s,r+1}$. The ordering of points according to values I_{sr} in Tables I and II differs, even though in case of tied importances the ordering in Table I is made consistent with values of p slightly larger than .5. Note that points tend to be more important later in the game, and when the score is close, and more important when the server is behind than ahead, because he probably will win if he can pull even. For p = .5, however, symmetry forces the point (s,r) to be exactly as important as (r,s); i.e., whether the server is behind or ahead does not affect the importance. The range of importances opens up considerably as p moves from .5 to .6. The outcome of the least important point 3-0 alters the probability of winning by only .0492 if p = .6, while if p = .5 it has a .125 influence. With importances of 2-3 being .6923 percent for p = .6 and .50 for p = .5, it follows that the ratio of importances of the most and least important points increases rapidly as p gets larger $(I_{23}/I_{30} = p^2/q^4$ for p $\geq$ .5).

The concept of importance ordinarily would be utilized by choosing tactics appropriately for each score. The payoff then is the "time-importance"

$$T_{sr} = N_{sr} I_{sr}$$

which weights the importance I_{sr} by the expected number of times N_{sr} that the point (s,r) is played in one game. The initial point 0-0, although not especially important, gains considerably in time-importance because it occurs so frequently.

Time-importance has the following property. Suppose a server, who ordinarily has probability p of winning a point on his serve, decides that he will try harder every time the point (s,r) occurs. If by doing so he is able to raise his probability from p to $p + \varepsilon$ ($\varepsilon > 0$ but small) for that point alone then he raises his probability of winning the game from P_{00} (the probability of winning the game at the outset) to $P_{00} + \varepsilon T_{sr}$.

For example, if a server with p = .60 (Table II) singles out the point 2-3 for extra effort, and is able to increase p to .61 on this point, then he will raise his probability of winning the game from P_{00} = .7357 (column 3) to .7357 + .01 × .3067 = .7388 (column 8). This extra effort is required in 6.83% of the points (column 5), about once in 15 service points. We shall see that, against an equal opponent, this slight gain increases the probability of winning a three-set match from 50% to 51.6% (if a similar extra effort also is made at 2-3 when the opponent serves).

The sum of the T_{sr} values in column 8 is the "magnification factor" M of the game, with the following meaning. If the probability p of winning every point is increased to $p + \epsilon$, the probability of winning the game is increased from P_{00} to P_{00} + Mϵ (this is a good approximation for small ϵ). An equivalent expression is M = dP_{00}/dp. For example, with p = .5 then M = 2.5 from Table I and so a 1% advantage (51% chance of winning a point) is magnified through the play of an entire game into a 2.5% advantage for winning the game (52.5% chance).

The values G_{sr}, the "gains", in column 9 are obtained by accumulating all entries T_{sr} in column 8 to the left and above the corresponding column 9 entry and then dividing this total by M. For example, from Table II, M = 2.086, G_{22} = $(T_{23} + T_{22})$/M = (.3067 + .1595)/2.086. G_{sr} is used for the vertical axis of the Lorenz curve, Figure 1. It represents the portion of the magnification factor that can be gained by increasing effort in all points at least as important as (s,r). Thus MG_{sr} = dP_{00}/dp, and P_{00} is increased approximately to P_{00} + M$G_{sr}\epsilon$ if p is increased to p + ϵ only on those points at or above (s,r) in the table. The server need make this increased effort the fraction F_{sr} of the time, and since $F_{sr} < G_{sr}$, he gets the disproportionate "cumulative efficiency" for concentrating on all points as important as (s,r) of E_{sr} = G_{sr}/F_{sr} (column 10).

The emphasis on the server in the preceding language is misleading. As stated earlier, each point is exactly as important to the receiver, and if he can decrease the server's success rate from p to p - ϵ on the important points, over the same fraction of points, then he also will achieve the same gain G_{sr}.

The Lorenz curve (so-called because it was used first in 1905 by M.C. Lorenz in studying economic inequalities) of Figure 1 plots the gain G_{sr} against the fraction of points F_{sr} required to achieve the gain. The 45° line G = F would obtain if all points were equally important. The vertical distance between the 45° line and the Lorenz curve, G-F, measures the extra probability of winning the game gained by concentrating the extra effort on the important points, rather than to a haphazard portion F of the points. The area between the line G = F and the Lorenz curve is an accepted index of the inequality of importances, known as (one half of) the Gini coefficient, and increases as p increases. E_{sr} is the slope of the line connecting the origin (0,0) to (F_{sr}, G_{sr}).

For example, if p = .6, either player can achieve about three-fifths (exactly G = .6038) of the gain provided by increased effort in all points by increasing effort in about three-sevenths (exactly F = .4321) of the points. These points are 2-3, 2-2, 3-3, 1-2, 1-3, 0-2, 0-1, and 1-1. The overall efficiency of this allocation of increased effort is E_{11} = 1.397

A player might consider trying harder on the important points and resting on the unimportant ones. If he increased p from .60 to .61 on

half his service points, and decreased from .60 to .59 on the unimportant
half, he would increase his winning percentage by 0.0075 from 0.7357 to
0.7432. The gain is G = 0.6792 at F = 0.5000, by interpolating from
Table II. (To carry out this strategy exactly, the player must try harder
77% of the time at 3-2 and rest the other 23%). Then 0.0075 = .01 × 2.086
× 0.6792 - .01 × 2.086 × (1 - .6792). We shall see that repeating this in
all games against an equal opponent (both when serving and when receiving)
increases the probability of winning the match from 0.500 to 0.539.

The final column of Tables I and II is the "efficiency of effort" e_{sr}
for the point. It is calculated as the derivative dG/dF of the Lorenz
curve, or more conveniently from the importance as

$$e_{sr} = I_{sr} \times L/M$$

Efficiency of a point e_{sr} differs from cumulative efficiency E_{sr} in that
it involves only one point, rather than the average of many. In fact, E_{sr}
is the weighted average of all values e at least as important as (s,r),
weighted by their relative frequencies, f. The cumulative efficiency of
all points is unity, so the pointwise efficiency e_{sr} expresses the importance of (s,r) relative to the average value. Said differently,

$$\overline{I} = M/L$$

is the "average importance" of a point. It is .3704 for p = .5 and .3217
for p = .6. This is one reason why matches are less exciting when strong
servers are involved. It also partially explains recent changes in the
balls and surfaces to slow the game down and thereby reduce the server's
advantage. (Of course, the other reason has been to lengthen rallies.)

For p = .6, 2-3 is 2.152 times as important as the average point
(column 11), and 1-1, the eighth most important point, is the least important point which is more efficient than average (1.033). At this position
on the Lorenz curve, G-F is maximized. The initial point 0-0 ranks only
tenth in importance, and has less than average efficiency (0.826).

A third Lorenz curve is plotted in Figure 1, for p = .75, to show
that importance becomes more varied as p increases, and that the important
points occur even less frequently (65% of the importance is confined to
26% of all points played).

Importance of Sets, Games, and Points to Winning the Match. Suppose
two players are competing in a best-of-three-sets match, and each has the
same probability p of winning a point on serve, so that the players are
equal although p > .5. If one player increases his probability of winning
the important points, each by a small amount ε, then he increases his
probability of winning each game by the amount GεM, whether serving or
receiving. Here G is the gain for the points associated with increased
probability, and M is the magnification factor for the game.

Extending this result to the probability, P_M of winning the match
obviously depends on numerous details, but the result is surprisingly stable:
P_M is increased from 0.5 to about 0.5 + 11 Gε for a three-set match, and to
almost 0.5 + 14 Gε for a five-set match, for all values of p between .3 and
.7. For example, if p = .6 and increased effort is made on the most important points, corresponding to 43 percent of the time, resulting in ε = .01

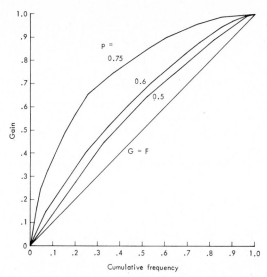

Fig. 1 Lorenz curves showing the gain G (fraction of the magnification M achieved) in probability of winning one service game as a function of the cumulative frequency F of improved play allocated to the most important points for p = 0.5 (Table I), p = 0.6 (Table II), and p = 0.75.

(a 61% chance of holding an important service point, a 41% chance of winning a point as receiver), then G = 0.6037 (from Table II) and so P_M = 0.566 is the approximate new probability of winning a three-set match.

Table III summarizes these results for sets and matches. The first three columns contain values already defined for games. The fourth column indicates the magnification factor M_{GS} resulting from a small increase in the probability of winning each game (serving and receiving) in a set (not stopped by a tie breaker). The probability of winning the set in the example p = .6 increases from .5 to $.5 + \epsilon M_{GS} = .5 + .01 \times 3.475 = 0.535$ if one player can find a way to raise P_{00} to 0.746 on his serve, and lower it to 0.726 for his opponent. The value M_{SM} = 1.50 is the derivative of $P^2 + 2P^2(1-P)$ (which is the probability of winning in the best of three sets given that P is the probability of winning one set) at P = .5. In the final column, M_{PM}, the magnification factor from point to match, is given as the product of M, M_{GS} and M_{SM}. It is close to 11 for the commonly experienced values of p. In a best-of-five set match, M_{SM} = 1.875 is calculated by differentiating $P^3 + 3P^3(1-P) + 6P^3(1-P)^2$, the probability of winning at least three of five sets, at P = 0.5. Replacing M_{SM} by this value increases each result in the rightmost column of Table III by a factor of 1.25.

We briefly consider the issue of other importances. It is easy to see that for equal players, each of the first two sets in a three-set match has importance .5, and that the third set (if there is one) has importance 1.0. Importances of games are more difficult to compute, and we merely summarize the results. For players of equal strength, the Lorenz curve is

Table III

Magnification Factors for Two Equally Matched Players

Probability Server Wins Point	Probability Server Wins Game	Magnification Factor, Points to Games	Magnification Factor, Games to Sets	Magnification Factor, Sets to Match	Magnification Points to Match
p	P_{00}	$M = M_{PG}$	M_{GS}	M_{SM}	M_{PM}
.50	.500	2.500	2.953	1.50	11.07
.55	.623	2.390	3.073	1.50	11.02
.60	.736	2.086	3.475	1.50	10.87
.65	.830	1.657	4.341	1.50	10.79
.70	.901	1.189	6.219	1.50	11.09

quite insensitive to the server's win probability p; the curves for $p = .5$ ($P_{00} = .5$) and $p = .6$ ($P_{00} = .736$), would nearly overlap one another and would be placed approximately midway between the curves for $p = .5$ and $p = .6$ in Figure 1. The most important games, as the reader should suspect by now, occur when the set is close and near the end of a set (4-4, 5-4, etc.), and the least important occur at highly lopsided scores such as 5-0 and 0-5. The ordering of the important games depends slightly on p; 5-3, 4-2, 3-1 and 2-0 all increase in importance as p increases.

The ultimate importance of winning a point is in relation to winning the match. The importance of a point to winning the match, I_{PM}, can be obtained from the importance of the point to winning the game I_{PG} (considered in the preceding section), from the importance of the game to winning the set I_{GS}, and from the importance of the set to winning the match I_{SM} (= 0.5 in the first two sets, 1.0 in the third set). We now state the

MULTIPLICATION RESULT: <u>For any point of any game of any set,</u>

$$I_{PM} = I_{PG} \times I_{GS} \times I_{SM}$$

This result follows directly from the definitions of importance and of conditional probability. Obviously the most important point in a traditional match is 2-3 (points), 4-5 (games), 1-1 (sets). It is match point against the server, but if he wins this point, then for $p = 0.6$ he has a 0.6923 chance of getting to 5-5, and a 50 percent chance of winning thereafter. The importance $I_{PM} = 0.346$ in this case. (With the 9-point tie breaker, the ninth point of the tie breaker in the final set has unit importance, $I_{PM} = 1.$)

Conclusion. Two additional interesting facts follow from the preceding theory. First, even though more points are served into the deuce court than into the ad court, the same total time-importance (T_{sr}) is associated with both sides of the court; hence, higher average importance is experienced in the ad court. This is one reason that doubles teams are advised to have the more experienced or stronger player receive on the left side. Second, the total time-importance associated with even-numbered and odd-numbered

service games are equal. Since there are more odd-numbered service games
the player who serves first generally will serve under less pressure.

In conclusion, the reader is cautioned not to implement this theory by
taking it too easy during the unimportant points. If a player doesn't try
on the unimportant points in his matches, there never will be any important
ones.

Acknowledgment. The author thanks Robert E. Machol for his considerable
editorial help and for his substantive suggestions.

Appendix

Given $0 \leq p \leq 1$ as the server's probability of winning a point, let
$q = 1 - p$, $t = pq$, $D = pq/(p^2 + q^2) = t/(1 - 2t)$. Let $T = p^2/(p^2 + q^2)$ be
the probability that the server wins the game given that the game reaches
deuce, and $\overline{T} = 1 - T = q^2/(p^2 + q^2)$ be the receiver's probability. We have
$P_{33} = T$ and

$$P_{sr} = pP_{s+1,r} + qP_{s,r+1} \tag{A.1}$$

with obvious definitions of zero and unit probability when $r = 4$ or $s = 4$
and $|s-r| \geq 2$. It follows for $r + s \leq 4$ that

$$I_{sr} = pI_{s+1,r} + qI_{s,r+1} \tag{A.2}$$

interpreting the importance as zero if the game is over, and using $I_{32} = \overline{T}$,
$I_{23} = T$, $I_{33} = D$.

While A.1 and A.2 provide convenient methods for evaluation of these
quantities on a computer, Table A.1 below contains closed-form expressions
sometimes more useful for theoretical work. Note that P_{sr} is obtained by
interchanging p and q in P_{rs} and subtracting this quantity from unity, i.e.,

$$P_{sr}(p) = 1 - P_{rs}(q)$$

and we use this to save writing out the cases $s > r$ in Table A.1. The
probability of winning the game is

$$P_{00}(p) = p^2 T(1 + 2q + 4q^2 + 8q^3) = \frac{p^4 - (2t)^4}{p^4 - q^4} \quad \text{if } p \neq 0.5 \tag{A.3}$$

with the elegant expression A.3 due to Kemeny and Snell [1960]. To save
space in Table A.1, we note here that

$$P_{01}(p) = p^2 T(1 + 2q + 4q^2 - 2q^3)$$

Finally, the formula for the magnification M is

$$M = \frac{dP_{00}(p)}{dp} = 20 \ t \ (3 - 4t)D^2$$

and the expected length of a game is

$$L = 4 + 4t + 8t^2 + 40t^2 D$$

Table A.1

Formulas for Win Probabilities P_{sr}, Importance I_{sr}, and the Expected Number of Times N_{sr} that (s, r) is Played

Server's Score (s)		Receiver's Score (r)			
		0	1	2	3
0	P_{0r}	$P_{00}(p)$	$P_{01}(p)$	$p^2(1+2t)T$	$p^3 T$
	I_{0r}	$10\ t^2 D$	$2p(2+p)tD$	$p^2(1+2p)D$	$p^2 T$
	N_{0r}	1	q	q^2	q^3
1	P_{1r}	$1 - P_{01}(q)$	$p(1+q+2q^2)T$	$p(1+t)T$	$p^2 T$
	I_{1r}	$2q(2+q)tD$	$3tD$	$p(1+p)D$	pT
	N_{1r}	p	$2pq$	$3pq^2$	$4pq^3$
2	P_{2r}	$1 - P_{02}(q)$	$1-q(1+t)\bar{T}$	T	pT
	I_{2r}	$q^2(1+2q)D$	$q(1+q)D$	D	T
	N_{2r}	p^2	$3p^2 q$	$6p^2 q^2$	$10qtD$
3	P_{3r}	$1 - q^3\bar{T}$	$1 - q^2\bar{T}$	$1 - q\bar{T}$	T
	I_{3r}	$q^2\bar{T}$	$q\bar{T}$	$\bar{T}$	D
	N_{3r}	p^3	$4p^3 q$	$10ptD$	$20t^2 D$

A MODEL OF THE USGA HANDICAP SYSTEM AND
"FAIRNESS" OF MEDAL AND MATCH PLAY

by

Stephen M. Pollock

Based on an article in Operations Research, 1974

Introduction. Handicapping systems are customarily assumed to be devices by which games of skill between two (or more) unequally skilled competitors can be made more "interesting". That is, the chances that any one
particular competitor will win are somehow adjusted to be closer to each
other than they would have been if there were no handicap system. The
sports of golf, bowling, sailboat racing, and horse racing have evolved
such "equalizing" schemes. The purpose of this chapter is to analyse how
well one particular handicapping scheme works: that used by the United
States Golf Association to make the game of golf attractive even when a
duffer goes head-to-head with a 5-handicap opponent. In the discussion
that follows, it will be assumed that the competition is between two players, playing a single 18-hole competition. Extension to 72-holes, tournaments, more than two players, etc. is straightforward.

In order to analyze the effect of this handicapping system, and to see
how it makes a particular contest "fairer", it is an advantage to develop
a purely mathematical model of the game of golf, as well as the system of
handicapping. Once this model has been established, involving assumptions
about the game's structure, the capabilities of the players, and the course,
we can answer the following questions:

What is the probability of player A beating player B
at medal play?

What is the probability of A beating B at match play?

(The distinction between medal and match play is in the scoring scheme: in
medal play, a player's total score on 18 holes is the determining factor;
in match play each individual hole is won or lost, and the total number of
holes won by a player is the determining factor.)

The Basic Model. As with all mathematical models of real phenomena,
a number of approximations and assumptions must be made so that sufficient
exercising of the model will enable us to understand the system being mod-
eled. It is not claimed that the real game of golf is completely captured
by descriptions which follow. Rather, the game is sufficiently close to
the model so that the insights to be gained are qualitatively--and to some
extent quantitatively--valid, and useful for decision making.

For two players, A and B, it is assumed:

1. A player is completely characterized by a probability
    distribution over the number of strokes that will be
    taken on each of 18 holes of a course.

2. This distribution is stationary (i.e. does not change
    with time).

141

3. Hole-to-hole scores are independent.

4. A player's individual hole scores are independent
    of the other player's score. The last two are per-
    haps our weakest assumptions, particularly for
    match play. They may be relaxed, however, in a
    straightforward but tedious way, by considering
    both A's and B's scores on successive holes to be
    statistically dependent and hence jointly distri-
    buted.

The effect of relaxing these assumptions is discussed qualitatively below.
The non-mathematical reader can skip to the numerical results to see the
implications of these assumptions.

We now can describe the probabilistic outcomes of competition under
four distinct conditions: medal or match play, and with or without use of
the handicapping system.

Medal Play Without Handicap. Let the number of strokes on hole i
($i=1,2,\ldots,18$) for player A be X_i, and for player B be Y_i, where X_i and Y_i
are random variables. Their respective probability mass functions are:

$$p_i(x) = \text{prob}\{X_i = x\} = \text{prob}\{A \text{ takes } x \text{ strokes on hole } i\},$$

$$q_i(y) = \text{prob}\{Y_i = y\} = \text{prob}\{B \text{ takes } y \text{ strokes on hole } i\}.$$

If we define $X = \sum_{i=1}^{i=18} X_i$ and $Y = \sum_{i=1}^{i=18} Y_i$ to be the total medal-play raw
scores for A and B respectively, then X and Y are random variables with
probability mass functions given by:

$$p(x) = \text{prob}\{X = x\} = p_1(\cdot)*p_2(\cdot)*\ldots*p_{18}(\cdot), \tag{1}$$

$$= \text{prob}\{A \text{ takes } x \text{ strokes on all 18 holes}\}$$

$$q(y) = \text{prob}\{Y = y\} = q_1(\cdot)*q_2(\cdot)*\ldots*q_{18}(\cdot). \tag{2}$$

$$= \text{prob}\{B \text{ takes } y \text{ strokes on all 18 holes}\}$$

where the $*$ indicates the convolution operator, which, for $f(\cdot)$ and $g(\cdot)$
defined to be probability mass functions on $(0,1,2,\ldots)$, is given by

$$f(\cdot)*g(\cdot) \equiv \sum_{z=0}^{z=\infty} f(z)g(x-z).$$

With these probability functions defined, it follows that

$$W_A = \text{prob}\{A \text{ wins the game}\} = \text{prob}\{X < Y\} = \sum_{y=1}^{y=\infty} \sum_{x=0}^{y-1} p(x)q(y), \tag{3}$$

$$W_B = \text{prob}\{B \text{ wins the game}\} = \text{prob}\{X > Y\} = \sum_{y=0}^{y=\infty} \sum_{x=y+1}^{x=\infty} p(x)q(y), \tag{4}$$

$$t = \text{prob}\{A \text{ and } B \text{ tie}\} = \text{prob}\{X=Y\} = \sum_{x=0}^{x=\infty} p(x)q(x). \tag{5}$$

This is, strictly speaking, a complete mathematical description of the competition, but without the data (or other information) to assess $p_i(x)$ and $q_i(x)$, the calculation of W_A and W_B is impossible. Furthermore, we have assumed that ties are possible, that is, the value of t calculated by equation (5) might not be zero. There are three ways to look at ties, each of which is consistent with the model and with actual golfing behavior.

1. Ties can be allowable outcomes of competition. Indeed, it can be argued that one purpose of the handicapping system is to increase the value of t.

2. Competitions ending in ties are considered to be replayed. In this case, due to basic assumptions 2 and 3, the recomputation of W_A and W_B will again be by the use of equations (3) and (4), and we see that

$$W_A' = \text{prob}\{A \text{ eventually wins, where ties are replayed}\}$$

$$= W_A/(W_A+W_B) = W_A/(1-t)$$

3. The probability descriptions $p(x)$ and $q(x)$, obtained from equations (1) and (2), are close enough to being continuous so that $t \approx 0$. This will, in fact, be the case when additional assumptions are made below. In this case, the handicapping system serves not to increase t, but simply to equalize the values of W_A and W_B.

Medal Play with Handicap. Now we suppose that players A and B have handicaps of h_A and h_B (the calculation of these handicaps is discussed below). The handicap difference allowable to A is defined to be $\delta \equiv h_A - h_B$. The handicapping system requires that δ strokes be subtracted from player A's score before comparing it with player B's score. From this it follows that

$$W_A = \text{prob}\{X-\delta<Y\} = \sum_{y=1-\delta}^{y=\infty} \sum_{x=0}^{x=y+\delta-1} p(x)q(y),$$

$$W_B = \text{prob}\{X-\delta>Y\} = \sum_{y=0}^{y=\infty} \sum_{x=y+1+\delta}^{x=\infty} p(x)q(y),$$

$$t = \text{prob}\{X-\delta=Y\} = \sum_{x=\delta}^{x=\infty} p(x)q(x-\delta).$$

Although the specific quantitative effects of introducing a handicap differential δ are still masked by the details of these equations, we can see qualitatively that as δ increases, W_A increases.

Match Play without Handicap. If a game is played hole by hole, such that a hole is 'won' on the basis of the score on that hole, then it is possible to calculate $a_i = \text{prob}\{A \text{ wins on hole } i\}$ and $b_i = \text{prob}\{B \text{ wins on hole } i\}$:

$$a_i = \sum_{y=1}^{y=\infty} \sum_{x=0}^{x=y-1} p_i(x)q_i(y),$$

$$b_i = \sum_{y=0}^{y=\infty} \sum_{x=y+1}^{x=\infty} p_i(x)q_i(y),$$

and

$$t_i = \text{prob}\{A \text{ and } B \text{ tie on hole } i\} = \sum_{x=0}^{x=\infty} p_i(x) q_i(x).$$

We now may define a new random variable Z_i

$$Z_i = \begin{cases} -1 \text{ if B wins hole } i, \\ 0 \text{ if A and B tie hole } i, \\ 1 \text{ if A wins hole } i. \end{cases}$$

If the match is played out to the full 18 holes (as we shall assume), then

$$Z \equiv \sum_{i=1}^{18} Z_i \tag{6}$$

is the number of holes by which A wins the match (Z is negative if B wins).

Thus, in match play, the win and tie probabilities may be calculated from $f(z) \equiv \text{prob}\{Z=z\}$*, since it is easily seen that

$$W_A = \sum_{z=1}^{18} f(z) \tag{7}$$

$$W_B = \sum_{z=-18}^{-1} f(z)$$

$$t = f(0)$$

The same comments on tie results hold as in the medal-play case.

Match Play with Handicap. The handicap system is more complicated when applied to match play as compared to medal play. Each hole is tradition-ally assigned a handicap number, indicating its relative difficulty; for details of this assignment, somewhat arbitrary in both theory and practice, see p. 155. When a handicap difference of δ is allowed to A in match play, A is allowed 1 stroke off each of the first δ holes ordered by handicap number. We may assume without losing generality that the holes have al-ready been ordered according to their handicap number, so that hole 1 is the most difficult, etc. To take handicaps into account, the calculations proceed exactly as before except for $i = 1, 2, \ldots, \delta$, in which case the indi-vidual hole probabilities are modified to be

$$a_i = \sum_{y=0}^{y=\infty} \sum_{x=0}^{x=y} p_i(x) q_i(y),$$

$$b_i = \sum_{y=0}^{y=\infty} \sum_{x=y+2}^{x=\infty} p_i(x) q_i(y),$$

$$t_i = \sum_{x=1}^{x=\infty} p_i(x) q_i(x-1).$$

---

*It can be shown that $f(y)$ has the discrete transform $\prod_{i=1}^{18} \left(\dfrac{b_i}{s} + t_i + a_i s \right)$

Calculation of Handicap. The current USGA regulations define a "handicap differential" to be the difference between par and a player's raw score in an 18-hole game. Thus, if $X(k)$ is a player's score on the kth most recent game (k=1 refers to his last game), then the differential $D(k)$ is $D(k) = X(k) - \rho(k)$, where $\rho(k) \equiv$ "par" for the course of the kth most recent game. As with the assignment of handicap numbers to specific holes, the assignment of a par value to a course is a mixture of tradition, consensus, and experience: it is meant to summarize the relative difficulty of the course. For the purposes of our model, however, it is appropriate to assume that par has a single value, ρ, for all games.

The handicap is defined to be the average of the lowest 10 differentials out of the last 20 games, multiplied by a factor α (here assumed to be 0.85, but see page 151).

$$H = (\alpha/10)\Sigma D(k),$$

where the sum is over the 10 lowest values of $D(\cdot)$. (In practice H is then rounded off to the nearest integer. In the analysis that follows, this has not been taken into account. The overall effect, in particular with regard to handicap differences, is negligible.)

The handicap H is a random variable, but it is only of interest to calculate its expected value $h = E[H]$, since it is this value that characterizes any given player as far as long-run ability to achieve par is concerned.

The expected handicap (hereafter simply referred to as handicap) may be calculated from a player's probability distribution of game scores. In particular, for player A with $p(x) = \text{prob}\{\text{score}=x\}$, it can be shown (using the usual order-statistics formulas) that $h_A = \alpha(\gamma-\rho)$, where

$$\gamma = 2 \sum_{x=0}^{x=\infty} xp(x) \sum_{i=1}^{i=10} \binom{19}{i-1} P(x)^i [1-P(x)]^{19-i}, \tag{8}$$

$$\rho = par, \quad P(x) = \sum_{n=0}^{n=x} p(n).$$

Player B's handicap is of course calculated with $p(\cdot)$ replaced by $q(\cdot)$.

The mathematical description developed above is (in theory) sufficient for computing W_A, W_B, and t for any of the four cases of interest. All that is needed are the hole-by-hole distributions $p_i(x)$ for A and $q_i(y)$ for B, as well as values for α and ρ. In order to gain some qualitative insights, however, it is now convenient to make additional assumptions which will reduce the number of parameters needed to describe a competition fully, so that some numerical results may be obtained and studied.

Equivalence of Holes and Normality of Hole Scores. Suppose we assume that the means and variances of the scores on each hole are identical, and known for player A and player B:

$$E(X_i) = \mu_A, \quad V(X_i) = \sigma_A^2, \quad (i=1,2,\ldots,18)$$

$$E(Y_i) = \mu_B, \quad V(Y_i) = \sigma_B^2, \quad (i=1,2,\ldots,18)$$

Then, by the central limit theorem, $X = \sum_{i=1}^{i=18} X_i$ and $Y = \sum_{i=1}^{i=18} Y_i$ are roughly normally distributed with moments

$$E(X) = 18\mu_A, \qquad V(X) = 18\sigma_A^2, \qquad E(Y) = 18\mu_B, \qquad V(Y) = 18\sigma_B^2.$$

In this case, when the handicap difference is δ, the medal play result is

$$W_A = \text{prob}\{A \text{ wins}\} = \text{prob}\{X < Y + \delta\} = \text{prob}\{X - Y < \delta\}.$$

Since X and Y are independent, normally distributed random variables, then X-Y is also normally distributed, with

$$E(X-Y) = 18(\mu_A - \mu_B), \qquad V(X-Y) = 18(\sigma_A^2 + \sigma_B^2),$$

and so

$$W_A = \Phi\{[\delta - 18(\mu_A - \mu_B)]/\sqrt{18}\,(\sigma_A^2 + \sigma_B^2)^{\frac{1}{2}}\} = 1 - W_B, \qquad (9)$$

where $\Phi(w)$ is the unit normal integral $\Phi(w) = \int_{-\infty}^{w} (1/\sqrt{2\pi})\,e^{-x^2/2}\,dx$.

   With the approximation, the probability of a tie becomes zero which (as discussed above) is the equivalent of having tie games replayed.

   The calculation of W_A for match play proceeds in a similar way. First, the individual hole-winning probabilities must be calculated (as functions of μ_A, μ_B, σ_A^2, σ_B^2). In the numerical results that follow, it has been assumed that X_i and Y_i are (discrete) random variables with an approximately normal cumulative distribution function. Thus, for holes $1, 2, \ldots, \delta$ the probability that A wins can be approximated by

$$a = \Phi\{[1 - (\mu_A - \mu_B)]/(\sigma_A^2 + \sigma_B^2)^{\frac{1}{2}}\} = 1-b, \qquad (10)$$

and for holes $\delta+1, \delta+2, \ldots, 18$,

$$a' = \Phi[-\ (\mu_A - \mu_B)/(\sigma_A^2 + \sigma_B^2)^{\frac{1}{2}}] = 1-b'. \qquad (11)$$

   The random variable Z, defined in equation (6) is the sum of 18 random variables Z_i, and thus its distribution is also approximately normal. The mean and variance of Z can be calculated directly

$$E(Z) = 2[\delta a + (18-\delta)a'] - 18, \quad V(Z) = 4[\delta a(1-a) + (18-\delta)a'(1-a')].$$

By using these formulas and the normal approximation, (7) reduces to

$$W_A = \Phi\{[\delta a + (18-\delta)a' - 9]/[\delta a(1-a) + (18-\delta)a'(1-a')]^{\frac{1}{2}}\}. \qquad (12)$$

   Handicap Difference. We now need to relate the handicaps h_A and h_B to the parameters μ_A, μ_B, σ_A, σ_B more transparently than through equation (8). Although this expression seems formidable, it is possible to show that the average of the lowest N out of 2N random variables, identically distributed with pdf $p(x)$, becomes, as $N \to \infty$,

$$2 \int_{-\infty}^{x_{(0.5)}} xp(x) \, dx,$$

where $x_{(0.5)}$ is the median of $p(x)$. With the normal approximations we have been making, equation (8) reduces to

$$h_A = \alpha[18\mu_A - (6/\sqrt{\pi})\sigma_A - \rho]. \tag{13}$$

Using the similar result for h_B allows the calculation of δ:

$$\delta = h_A - h_B = \alpha[18(\mu_A - \mu_B) - (6/\sqrt{\pi})(\sigma_A - \sigma_B)]. \tag{14}$$

Numerical Results. Numerical results using these approximations are shown in Figs. 1-3, where W_A = probability {player A wins} is plotted against the handicap difference δ. These results were obtained by fixing values of μ_A, σ_A, μ_B, σ_B, and $\alpha = 0.85$, and then calculating δ from (14) and the probabilities from (9) and (12).

In interpreting these curves, it should be noted that

(a) Player A is assumed to be the weaker player on the average $(\mu_A \gtreqqless \mu_B)$

(b) The values of σ_A and σ_B represent single-hole standard deviations, and thus "consistency" of the players. The standard deviation for an 18-hole game is found by multiplying these values by $\sqrt{18} \approx 4.24$. Thus the figures show results for game standard deviations from about 2.5 to 6 strokes, representing a spectrum of golfers.

(c) The handicap system has two factors built into it that mitigate against 'fairness' (i.e., producing $W_A = 0.5$ for all cases). The first is the parameter α, which influences the slopes of the curves. The second is the "lowest 10 out of the last twenty" calculation, which allows σ_A and σ_B to influence the intercept on the W_A axis.

The curves show interesting but not surprising results. From Fig. 1, an inconsistent (high σ_A) player, A, playing a stronger opponent, B, should prefer match to medal player whether σ_B is high or low. The handicap system never increases W_A to 0.5, but prevents W_A from being disastrously small. Note that negative handicap differences are possible even though A is a worse player on the average. A moderately consistent A (Fig. 2) should still prefer match to medal play unless the opponent B (better on average) has high σ_B. In this case B may have obtained his low handicap by shooting some low scores and some high ones, and two effects become apparent:

(a) If the handicap difference is less than some given value ($\delta \cong 8$ for $\sigma_B = 1.4$, $\sigma_A = 1.0$), then A actually has a better than 0.5 probability of winning in either match or medal play.

(b) Below this value of δ, A has a better chance playing medal rather than match.

Finally, Fig. 3 shows the results for a high-handicap but consistent player. Depending upon the consistency of B, there is a value of the handicap difference such that, below this value, A prefers medal play to match play.

Relation between Consistency and Average Score. It is of obvious interest to determine what relation--if any--exists between μ and σ (it seems reasonable that a 'better' player would have low values for both). Some recent small sampling (600 rounds by about 50 players at one course) has indicated that, for values of handicaps between 2 and 20, the following relations 'bracket' the dependence between σ and h:

$$0.71 + 0.04h \leq \sigma \leq 0.85 + 0.02h,$$

Using these in equation (13), with $\alpha = 0.85$, gives

$$0.031(18\mu-\rho) + 0.63 \leq \sigma \leq 0.016(18\mu-\rho) + 0.89,$$

which bracket (roughly) the relation between σ and $(18\mu-\rho)$, which is the average number of strokes per game above par. These relations may be used to get a rough idea of a player's consistency, as expressed by σ, if a sufficient set of statistics have not been kept to calculate σ directly.

Relaxation of Assumptions. A number of assumptions have been made in order to reach the point where numerical results could be obtained. It is possible, however, to examine the qualitative effects produced by the relaxing (or changing) of these assumptions.

The stationarity of the competitors' performance is, of course, crucial. It is hard enough to characterize the game where the skill or conditions of play are constant in time. The dynamic effects of learning, seasonal changes, and aging are left to other analysts.

The dependence of hole-to-hole scores, and between opponents' scores on particular holes, are also quite real. In fact, the boldness or conservativeness of hole approaches, when involved in match play, is probably a major determinant of a particular player's value of σ. There seems to be no way to adjust the model presented above to account for this kind of adaptive play. However, if it can be assumed that the competitors are good enough (or bad enough) not to adapt their style of play to the other's score, but simply to the hole itself, then the results are still valid. This is true even if a given player has hole-to-hole dependence: he plays in good and bad "streaks". For with the probabilistic results we have been considering, the independence of the players will produce an averaging-out effect.

Finally, in the numerical examples, hole scores were assumed to be identical. This, in fact, is not necessary to get the result of equation (9). In the more general case, it is sufficient to use this equation, with the parameters now defined to be:

$$\mu_A \equiv \frac{1}{18} \sum_{i=1}^{18} \mu_A(i)$$

$$\sigma_A^2 = \frac{1}{18} \sum_{i=1}^{18} \sigma_A^2(i)$$

where $\mu_A(i)$ and $\sigma_A^2(i)$ are the ith-hole mean and variance. The only difference arises in equations (10) and (11), where it is possible (indeed likely) that $\mu_A(i)$ and $\sigma_A(i)$ will be different for the low-handicap-number holes $(1,2,\ldots,\delta)$ than for the others. However, this effect will also occur for player B, and the total effect on the curves shown in the Figures would thus be second-order, except for highly mismatched opponents.

Acknowledgment. Appreciation is expressed to Professor W. Steffy, who introduced me to the issue of the fairness of medal vs. match play. R. Machol has also been helpful in the presentation of this chapter.

SUMMARY

This paper presents a representative model for the game of golf. Although many simplifying assumptions are made, the results offer general insights into the relations between handicaps and win probabilities for selected pairs of players. The model shows the direction and magnitude of the changes in winning probabilities due to (a) having a choice between match and medal play, and (b) the parameters and calculations involved in the current handicap scheme. In addition, the model points out the importance of considering not only the average performance of a player (reflected by the parameter μ) but also the player's consistency (represented by σ). The fundamental results allow a great deal of room for numerical exploitation. In particular, it is possible to use the model to explore the effect of changing the parameter α in the handicap computation, and altering the "lowest ten out of the last twenty" averaging rule.

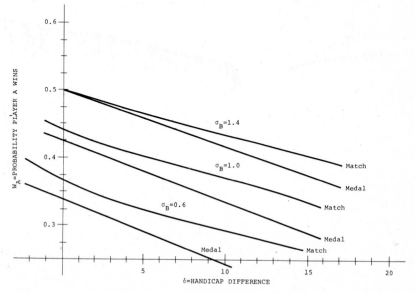

Figure 1. W_A vs.   For $\sigma_A = 1.4$

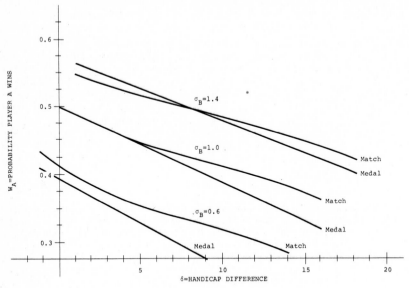

Figure 2. W_A vs.   For σ_A = 1.0

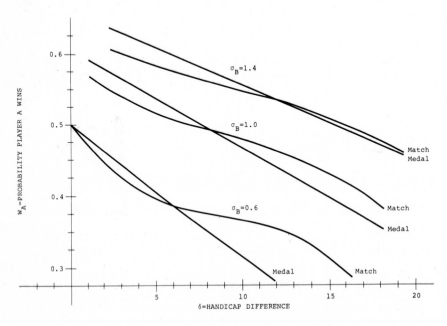

Figure 3. W_A vs.   For σ_A = 0.6

AN EVALUATION OF THE HANDICAP SYSTEM OF THE UNITED STATES GOLF ASSOCIATION

by

Francis Scheid

Based on articles in <u>Golf Digest</u>,[*] 1971 and 1973,
and <u>SIAM Journal of Applied Mathematics</u>,[**] 1972

<u>Introduction</u>. This paper summarizes a study which used thousands of actual golf scores and found: (1) that the present USGA handicapping system underestimates differences in playing ability and puts weaker players at a significant disadvantage in competition, (2) that the system rates the normal golf course quite accurately but makes errors worth correcting on unusually long or short courses, particularly par threes, and (3) that there is a serious need to adjust handicaps earned on one course for fair play at another.

The system rates courses by the formula $R = .005Y + 38.25$ where Y is the total length in yards, and then allows small modifications for unusual circumstances. A recent change reduces ratings of short courses by .001 stroke for each yard under 6000. It then assigns each player a handicap by computing the average amount by which the ten best of his most recent twenty scores exceed course rating and then multiplying by 96 percent. This system is in general use throughout the USA as well as in many other countries. (The factor .96 was made effective on Jan. 1, 1976, as a result of this study, the previous factor having been .85.)

<u>Simulated play.</u> In two-player competition the USGA recommends that the stronger allow the weaker to subtract from his score a number of strokes equal to the difference in their handicaps. At stroke play it is the total for the round which is reduced, while at match play strokes are given at designated holes. How many should be given to make a fair match? To answer this question, scorecards were collected first at the Plymouth Country Club in Plymouth, Massachusetts, until twenty rounds for each of fifty golfers were in hand. This made it possible to play 400 simulated matches by computer for any pair of golfers, simply by comparing rounds, and so find the number of strokes actually needed on the average. All together about two million matches were "played". The result for match play is shown as Fig. 1. For instance, at handicap difference of 5 the stronger player won 67, 61, 53, and 47 percent of the time depending on whether 4, 5, 6, or 7 strokes were given. This suggests that the official recommendation leaves the odds 61-39 in favor of the better player and that between 6 and 7 strokes are needed to reduce the odds to 50-50. The other curves tell similar stories; in no case was the full handicap difference enough.

<u>Smoothing</u>. The curves of Fig. 1 are more or less parallel and equally spaced. With a natural ceiling and floor at the 100 and 0 levels, and anticipating that they should be antisymmetric about the 50 level, a family of cubics of the form

$$y = 50 + s(x-hi) + t(x-hi)^3$$

*Parts of this article through the courtesy of <u>Golf Digest</u> magazine, copyright 1971 and 1973.

**Parts of this article copyright 1972 by Society for Industrial and Applied Mathematics.

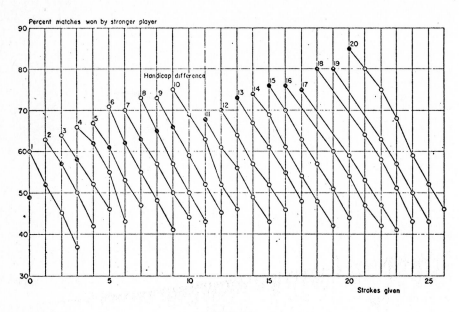

FIGURE 1

Two-man match play (raw data).
Solid circles represent USGA recommendation.
Handicap differences are those in use at the
time of the experiment, with the factor .85.

seems reasonable for smoothing the data. Here x represents strokes given,
y winning percentage of the stronger players, and i handicap difference.
Least-squares fitting to the simulation data produced

$$y = 50 - 6.95(x - 1.27i) + .03(x - 1.27i)^3$$

which predicts values differing by an average of only 1.3% from the raw experimental data.

Results for two player events. To equalize the players y must be 50,
and by the cubic formula this happens when x = 1.27i. This result induced
the USGA to authorize a corroborating study and Bogevold [1974] using scores
from various parts of the country found the identical figure. The subsequent shift to the .96 factor achieved part of the needed 27% inflation and
the above result can now be stated as

strokes needed = 1.12(new handicap difference).

If this ideal number of strokes is not given our cubic predicts the winning
probability of the stronger player. Certain values of special interest are
displayed in Table I. Even a quick glance suggests that by offering less
than full difference, as is customary, low-handicappers are showing more
greed than courage.

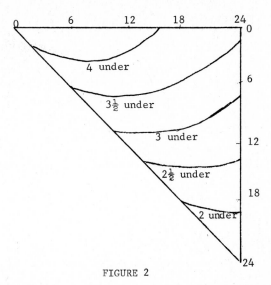

FIGURE 2

Better-ball of two-man teams, full handicaps allowed,
smoothed result. Handicaps of partners are along upper and right edges.

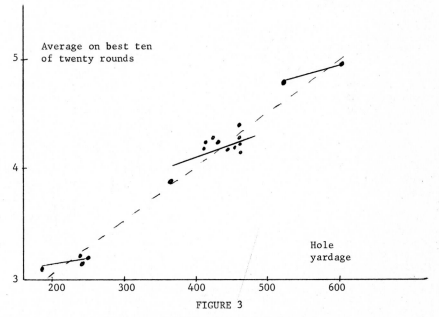

FIGURE 3

Increase of difficulty with hole yardage.

Corresponding results for stroke play were almost identical, the number of strokes needed being slightly greater and the odds in favor of the stronger player rising only slightly faster when the proper number is not given.

New handicap difference	1	2	3	4	5	6	7	8	9	10	11	12	13	14	15
One stroke less	.58	.59	.59	.60	.61	.62	.63	.63	.64	.65	.66	.67	.67	.68	.69
Full handicap difference	.51	.52	.53	.53	.54	.55	.56	.57	.58	.58	.59	.60	.61	.62	.62
One stroke more	.44	.45	.46	.46	.47	.48	.49	.50	.51	.51	.52	.53	.54	.55	.56
Two strokes more									.44	.45	.45	.46	.47	.48	.49

Table I.   Probability that stronger player wins,
two-man match play, after smoothing.

Better-ball play. At fourball and member-guest types of competition, the better ball of a two-player team counts. To determine how it varies with handicap and strokes given, several hundred teams were formed and the usual 400 rounds simulated for each pair. This was done at scratch play (no strokes given) and for various percentages of handicap. Figure 2 shows one of the main results after smoothing. Reading the handicap of the stronger player at the right and that of the weaker at the top the average better-ball can be estimated from the contour curves provided. The main thing to notice is, of course, that even with 100% of handicaps given, the advantage is still with the stronger players: they manage a score about two strokes better than the duffers at the lower right. Only at 107% was near equality achieved, and beyond this the duffers got the edge. Another point worth noting is that the common form of team "balancing", in which handicap totals of partners are equalized, is only roughly successful. Teams along any NE-SW diagonal would be so balanced, but even at 100% show some score variation. Better balancing could be arranged by following the contour curves.

The recent change from .85 to .96 as factor in the handicap formula means that the ideal figure of 1.07 for better-ball play becomes .95 in terms of the new handicaps. It is interesting to observe that the USGA has just recommended .80 for this role, so the poor duffer is still to be denied his day.

Course ratings. The idea of scratch play as used in golf can be described as "the scratch principle", that is, the scratch golfer measures to handicap 0 on any golf course. This makes him the reference player for the system, other handicaps measuring differences in ability from scratch level. It also means that courses are rated the way scratch players perform on them, on the best ten of twenty rounds. In order to test consistency with the scratch principle and to study the performance of golfers at other levels, hole-by-hole data was obtained from both the PGA and the USGA. Computing averages (again best ten of twenty rounds) on holes of length varying from short par 3's to long 5's, a somewhat surprising result appeared. Figure 3 shows the picture for one touring pro and it is typical of golfers at all ability levels. The least-squares lines fit to the separate par categories are usually less steep than the overall regression line. This was true in 372 out of 601 cases tested, an event which would have probability .000001 if the overall line were a good fit. The analysis of variance test using the F distribution also rejects linearity at well beyond the 99% level of significance. In view of this nonlinearity, separate formulas of the type

$$P = (A+BH)Y + C+DH$$

were fitted to each par category, Y being hole yardage and H handicap. This form allows both intercept and slope to change with ability level. Least-squares fitting produced

Par 3 $P = (.0041 + .000043\ H)Y + 2.31 + .034\ H$

Par 4 $P = (.0042 + .000072\ H)Y + 2.43 + .034\ H$

Par 5 $P = (.0039 + .000050\ H)Y + 2.69 + .042\ H$

which become hole-rating formulas when H=0. When applied to more than 200 courses, these formulas showed that for courses of more or less normal length, ratings are reasonably accurate--the average absolute change being only .2 strokes. For other courses more important discrepancies occur. In fact, overall length is not a sufficient indicator of course difficulty; it is the particular combination of hole lengths that governs. Par 3 courses rate special mention, having the single rating formula $R = .0041Y + 41.6$ where Y is again total yardage. For a course of 3000 yards this gives 53.9 where the official USGA rating is now 50.3. This is an error of more than three strokes. A scratch player would measure to handicap 3.6, a serious violation of the scratch principle.

Handicap adjustment. When the above formulas are applied at other H levels a sharper view is obtained of another problem, the magnification of ability differences as courses get harder. Two extreme cases are shown at the top and bottom of Table II. The center row represents an average course, and therefore a standardized handicap; the others represent handicaps that would be computed locally. For a very weak player, the locally computed handicap must be reduced by four strokes at MONSTER and increased by eleven at PAR 3. A handicap 44 at the former really plays at the same ability level as a 29 at the latter.

The point is that with hole-rating formulas in hand a uniform handicapping system can be arranged, including all courses and ability levels. Clearly handicap adjustment between courses should be a basic part of it. The future of golf handicapping will require an adjustment table for each course.

MONSTER	-6	0	6	11	17	22	28	33	39	44
NORMAL	-5	0	5	10	15	20	25	30	35	40
PAR 3	-4	0	4	7	11	14	18	22	25	29

Table II. Local handicaps at three courses

THE OPTIMAL AIMING LINE

by

Shaul P. Ladany, John W. Humes, and Georghios P. Sphicas
Based on an article in Operational Research Quarterly, 1975

Introduction. In the running long jump, an athlete runs toward a
take-off line and leaps into the air. His individual score is the shortest
horizontal distance between the official take-off line and the rear portion
of his rear leg's foot-print in the landing pit, provided that at take-off
he does not foul (that is, the tip of his front "take-off" foot does not
extend forward of the official take-off line). The athlete is allowed
three jumps and his final score is the maximal recorded distance among the
three attempts.

In some cases, where a jumper has experienced a definite tendency to
foul, he will, in order to reduce his foul rate and thereby increase the
number of chances for selecting an optimal jump, aim at an imaginary line
behind the official take-off line. Thus, a jumper who is attempting to
gain the highest possible recorded distance in a competition is faced with
the decision, knowingly or not, to select the appropriate take-off aiming
line, according to his specific abilities.

Many papers have been written on various aspects of the running long
jump. Some have dealt with the techniques and mechanics of the jump, consisting of case studies of specific jumping techniques used to help develop the physical movements required for the event. Others have been
theoretical papers that use the laws of physics to yield quantitative information about the motion of the body during take-off and flight. Bunn
[1964], Cureton [1935], and Dyson [1963] have used equations of exterior
ballistics to describe the motion of the long-jump flight. Dittrich [1941]
measured take-off velocities by analyzing motion pictures and Ramey [1970]
determined the force relations in the long jump. No one, however, has
tried to investigate the probabilistic behavior of the take-off and its
implications. Brearley did consider it briefly--see his article.

In this paper an attempt is made to analyze the probabilistic behavior of this jumping distance and take-off accuracy and to determine the
most advantageous take-off aiming line according to an individual jumper's
abilities.

The Model. Assume that the real jumping distance, x, which is the
shortest horizontal distance between the tip of the forward shoe at take-
off and the aftermost contour of the rear leg's footprint after landing,
is a continuous random variable. Further, we may express the accuracy of
take-off as the distance, y, between the tip of the forward shoe at take-
off and the take-off line aimed at. This will result in a positive distance when "overshooting" and a negative one when "falling short" at take-
off.

Thus, when a jumper aims to take off from a line that is a distance,
a, behind the official take-off line, as shown in Figure 1, the distance d_i
which will be recorded for the jumper for each of his trials is

156

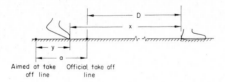

Figure 1

$$
d_i = \begin{cases} y_i + x_i - a & \text{for } y_i \le a, \\ 0 & \text{for } y_i > a. \end{cases} \tag{1}
$$

We are interested in the maximum scored distance, D, among the three trials which are assumed to be independent, i.e. $D = \max(d_1, d_2, d_3)$.

The expected value of D is obtained by summing up the probabilities of three mutually exclusive events: (1) the expected maximum score out of a single jump when two attempts have fouled; (2) the expected maximum score out of two jumps when one attempt has fouled; and (3) the expected maximum score out of three jumps when all the three attempts have succeeded. This is given by

$$
E(D) = \binom{3}{1} F(a)[1 - F(a)]^2 \, E[(x + y \,|\, y \le a) - a] \tag{2}
$$

$$
+ \binom{3}{2} F^2(a)[1 - F(a)] \, E\left\{ \max_{i=1,2} [(x_i + y_i \,|\, y \le a) - a] \right\}
$$

$$
+ \binom{3}{3} F^3(a) \, E\left\{ \max_{i=1,2,3} [(x_i + y_i \,|\, y \le a) - a] \right\}
$$

where $F(a) = P(y \le a)$.

The decision problem is to choose the value of a which maximizes E(D). For given distributions of x and y, the right-hand side of (2) could theoretically be expressed as a function of a single variable a. Unfortunately, analytical expressions are impossible to derive, even in the simple case of x_i and y_i being independent observations from normal distributions. The assumptions of normality and independence were tested by measuring x and y for 100 jumps by a single jumper in ten separate training sessions, using various values of a. This jumper had parameters $\bar{x} = 701.2$, $\sigma_x = 20.44$, $\bar{y} = .010$, $\sigma_y = 7.40$, all in centimeters. Chi-square tests showed that the hypotheses of normality and independence could not be rejected at the .50 level; the observed correlation coefficient between x and y was 0.11, not significantly different from zero. It follows that x + y is also normal and the moments of the extremes of identically normal observations are known. The difficulty is that $(x+y \,|\, y \le a)$ is the sum of a

normal and a truncated normal variable. The moments of extremes from such
a distribution are not easily derived, and if they were calculated they
would not allow an analytical solution to our problem. As alternative
approaches, we turn to simulation and to an approximation of (2).

   Simulation and Approximation. The simulation of the process for various values of a and then the selection of a_0, the value of a that provides
the maximal value of E(D), has the benefit that it allows also the deriva-
tion of the standard deviation of the distribution of E(D) as a simple by-
product.

   Let Z_a stand for the standard normal deviate, $Z_a = (a - \mu_y)/\sigma_y$. The results of the simulation using 3000 runs of three jumps for each value of Z_a
for the particular case observed are shown in Figure 2. It is apparent
that the simulation results are liable to random fluctuations as measured
by the low standard deviation observed for the expected maximum scored distance. Nevertheless, the tendency of the behavior of E(D) as a function
of Z_a is obvious and it points out that max E(D) is reached at about $Z_a = 1.2$
where it peaks to a value of about 707 cm.

   Since the simulation approach requires a separate series or runs for
each set of parameter values, an approximate solution for equation (2) was
attempted. In the specific case investigated, $Z_a \approx 1.20$, i.e. the right-
tail probability amounts to only 0.115. Therefore the probability distri-
bution of $x + y | y \leq a$ can be approximated by the distribution of x+y which
is normal, i.e. $k = (x+y) \sim N(\mu_x + \mu_y, \sigma_{x+y}^2)$, where $\sigma_{x+y}^2 = \sigma_x^2 + \sigma_y^2$. Thus

$$E\left[(x + y | y \leq a) - a\right] \approx \mu_x + \mu_y - a$$

The last part of the second term of equation (2) is approximated as follows:

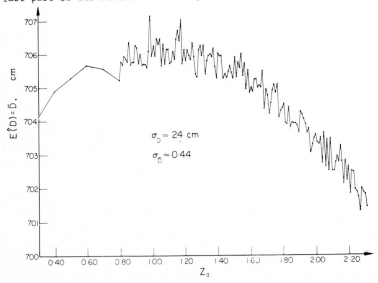

Figure 2

$$E(II) = E[\max\{[(x_1+y_1|y_1 \leq a) - a], [(x_2+y_2|y_2 \leq a) - a)]\}] \tag{4}$$

$$\approx E\{\max[(k_1 - a),(k_2 - a)]\}$$

$$= \int_{k=0}^{\infty} 2kF(k)f(k)\,dk - a \approx \int_{k=-\infty}^{\infty} 2kF(k)f(k)\,dk - a$$

where F(k) and f(k) are the cumulative distribution function, and the probability density function, respectively.

This integral is insolvable analytically. However, it has been calculated numerically by Tippett [1925] and graphed by Gumbel [1958] for standard N(0,1) distributions, providing a value of $0.564\sigma_{x+y} + \mu_x + \mu_y$. Therefore,

$$E(II) \approx 0.564\sigma_{x+y} + \mu_x + \mu_y - a \tag{5}$$

The last part of the third term of equation (3) is approximated similarly

$$E(III) = E\left[\max\{[(x_1+y_1|y_1 \leq a) - a],[(x_2+y_2|y_2 \leq a) - a],\right.$$

$$\left. [(x_3+y_3|y_3 \leq a) - a]\}\right]$$

$$\approx E\{\max[(k_1 - a),(k_2 - a),(k_3 - a)]\}$$

$$= \int_{k=0}^{\infty} 3kF^2(k)f(k)\,dk - a \approx \int_{k=-\infty}^{\infty} 3kF^2(k)f(k)\,dk - a$$

Neither is this integral solvable analytically, nor is it tabulated. However, Romanovsky [1933] links the successive means of maximums out of three identical standardized normal distributions to be 1.5 times the mean of maximums out of two identical standardized normal distributions. Therefore,

$$E(III) \approx \int_{k=-\infty}^{\infty} 3kF^2(k)f(k)\,dk - a = 0.846\sigma_{x+y} + \mu_x + \mu_y - a \tag{6}$$

Substitution of equations (4), (5), and (6) into (2) provides

$$E(D) \approx 3F(a)[1 - F(a)]^2(\mu_x + \mu_y - a) + 3F^2(a)[1 - F(a)]$$

$$[0.564\,\sigma_{x+y} + \mu_x + \mu_y - a]$$

$$+ F^3(a)[0.846\,\sigma_{x+y} + \mu_x + \mu_y - a]$$

and, after simplifying,

$$E(D) \approx [F^3(a) - 3F^2(a) + 3F(a)](\mu_x + \mu_y - a) + [2F^2(a) - F^3(a)](0.846\sigma_{x+y}) \tag{7}$$

It was attempted to derive from $\partial E(D)/\partial a = 0$ the optimal value of a, a_0, for which $E(D)$ is maximum. However, it was impossible to express a_0 (or the corresponding standard normal deviate Z_{a_0}) in closed form, and therefore a numerical search would have been required. Such a search has been applied instead directly to $E(D)$ of equation (7). The values of Z_{a_0} obtained for various combinations of μ_x, σ_y and σ_{x+y} are shown in Figure 3. The horizontal axis should be $\mu_x + \mu_y$, but since in most practical cases $\mu_y \approx 0$, it has been marked just as μ_x.

For the specific case of $\mu_x = 701.23$, $\mu_y = 0.0010$, $\sigma_x = 20.44$, $\sigma_y = 7.50$, $\sigma_{x+y} = 21.77$, Figure 3 provides an approximate optimal solution of $Z_{a_0} \approx 1.27$. Reference to the simulated results for this same case, Figure 2, shows that although $Z_a = 1.27$ does not appear to be optimal, the curve of $E(D)$ as a function of Z_a around the optimal region of Z_a is flat (insensitive to changes in Z_a). Thus $E(D)$ corresponding to $Z_a = 1.27$ is only slightly different from the max $E(D)$, so that in the present example the results of Figure 3 provide a valid approximation. Based on the above, it would be possible to use the general results of Figure 3 as workable approximations for the optimal aiming line.

Figure 3 provides information on the influence of the various parameters of Z_{a_0}: (1) Increase of μ_x slightly increases Z_{a_0}, but the influence of increased μ_x is bigger for lower σ_y; (2) Decrease in σ_y increases Z_{a_0}, but increase in Z_{a_0} is of such a magnitude that $a_0 = \mu_y + \sigma_y Z_{a_0}$ decreases; (3) Increase of σ_{x+y}, i.e. the increase of σ_x for a given σ_y, increases Z_{a_0} Furthermore, as the approximate value of Z_{a_0} of Figure 3 increases, the right-tail probability decreases and therefore the approximation error should be lower; consequently the reliability of the approximate results should be higher.

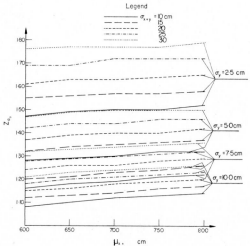

Figure 3

Conclusion. For the specific case it has been found that the optimal aiming line should be $a_0 \approx 1.20 z_{a_0} + \mu_y \approx 9$ cm, and the expected maximum jump will be 707 cm as opposed to (approximately) 620 cm if the official line is aimed at. For other cases the optimum can be read from the figures.

The maximization of expected length of jump is not the only criterion that could be pursued, nor would it fit the requirements of every athlete. As Lilien [1976] has pointed out, alternative criteria could be the maximization of the expected deviations of jumps above a certain non-zero minimum, or the maximization of the expected jumping utilities when a non-linear utility function might be introduced. The application of these criteria can be pursued in a similar manner to the one adopted in the paper, except that in equation (2) the expected maximum deviation or the expected maximum utility should be taken into account. Finally, the strategy need not be identical for all jumps: the first jump, like a first serve in tennis, might be incautious; the last jump might depend on previous jumps and on objectives. Dynamic policies are further explored by Sphicas and Ladany [1976].

The present model and the associated reasonings could be applied in an identical manner also to other track-and-field events such as the triple jump and javelin throw.

THE LONG JUMP MIRACLE OF MEXICO CITY

by

M. N. Brearley

Based on an article in Mathematics Magazine, 1972

Introduction. Did a miracle occur at the 1968 Olympic Games in Mexico City? The Games record for the long jump [Harlan, 1964], which had crept up since 1904 by five successive increments totalling 0.76 m, was raised by 0.80 m in a single mighty leap. In his first (and only) attempt, R. Beamon of the U.S.A. lifted the record from 8.10 m (26 ft 6 7/8 in.) to 8.90 m (29 ft 2 1/2 in.), an increase of 9.9%. The enormity of this feat may be gauged by noting that a similar improvement would cut a four-minute mile to 3:36.

It has been suggested that Beamon's leap owed much to the reduced air drag resulting from the lower air density at the high altitude (7400 ft) of Mexico City. We seek a mathematical model of the long jump which will enable the influence of the Mexico City altitude to be assessed quantitatively.

Setting up the Mathematical Model. The distance achieved in a long jump is affected by many factors such as spring off the board, the action of the free leg at take-off, rotation of the body by a hitch-kick, and landing technique. To model all such aspects would be difficult; fortunately it is unnecessary for the present purpose.

After take-off the athlete's center of mass G describes a path which can be investigated like that of any projectile moving through a resisting medium. For our purpose it will plainly suffice to compare the horizontal range of a projectile in vacuo with its range when similarly projected in a medium of known density. The effects of air densities at various altitudes can thus be compared without considering details of the long jump technique.

In traditional notation the equation of motion of the athlete after take-off is

$$m \, d\underset{\sim}{V}/dt = m\underset{\sim}{g} + \underset{\sim}{D} \tag{1}$$

where $\underset{\sim}{V}$ is the vector velocity of G at any instant and $\underset{\sim}{D}$ is the air drag on the athlete. The direction of $\underset{\sim}{D}$ is opposite to that of $\underset{\sim}{V}$, and its magnitude D is known from experimental work of Nonweiler [1956] to be given by

$$D = k\rho V^2$$

where ρ is the air density, $V = |\underset{\sim}{V}|$, and k is constant for a fixed body posture. Hence

$$\underset{\sim}{D} = -k\rho \underset{\sim}{\hat{V}}$$

where $\underset{\sim}{\hat{V}} = \underset{\sim}{V}/V$, the unit vector. On division by m, (1) becomes

$$d\underset{\sim}{V}/dt = \underset{\sim}{g} - KV^2\underset{\sim}{\hat{V}} \tag{2}$$

162

where

$$K = k\rho/m \tag{3}$$

It is convenient to refer the flight path of G to the traditional axes Ox, Oy, where O is the position of G at take-off. At time t after take-off, let

u,v = velocity components of G parallel to Ox, Oy.

The initial conditions of the flight are then

$$t = 0, \; x = y = 0, \; u = u_0, \; v = v_0 \tag{4}$$

where u_0, v_0 may for the present remain arbitrary.

The nonlinear differential equation (2) has no exact solution in closed form. Though it can be solved approximately to an accuracy sufficient for the present purpose, for the sake of brevity a less orthodox approach will be used. In this approach, all significant approximations will overestimate the effect of air resistance.

The Energy Dissipated by Air Resistance. The theory of projectiles in the absence of air resistance yields the familiar results

$$u = u_0 \qquad v = v_0 - gt \qquad y = v_0 t - \tfrac{1}{2}gt^2 \tag{5a,b,c}$$

$$v = \pm(v_0^2 - 2gy)^{\frac{1}{2}} \qquad h = \tfrac{1}{2}v_0^2/g \qquad R_0 = 2u_0 v_0/g \tag{6a,b,c}$$

where h and R_0 are the height and range attained relative to a horizontal plane. From the principle of conservation of energy it readily follows that

$$V = (V_0^2 - 2gy)^{\frac{1}{2}} \tag{7}$$

where

$$V_0^2 = u_0^2 + v_0^2 \tag{8}$$

In the presence of air resistance the energy per unit mass dissipated during the total flight time t_1 is

$$E = \int_0^{t_1} KV^2 \hat{\underline{v}} \cdot \underline{v} \, dt = \int KV^3 v^{-1} dy \tag{9}$$

where the latter integral is evaluated over the total domain of y. It may be verified that the change in velocity caused by air resistance is very small (averaging about 1.3% in a typical case), so that E may be found very accurately by using (6a) and (7) to approximate the integrand in (9). It is found that

$$E = \int_0^h KV^3 (v_0^2 - 2gy)^{-\frac{1}{2}} dy - \int_h^0 KV^3 (v_0^2 - 2gy)^{-\frac{1}{2}} dy$$

$$= 2K \int_0^h (V_0^2 - 2gy)^{3/2} (v_0^2 - 2gy)^{-\frac{1}{2}} dy$$

On changing to the new variable

$$z = (v_0^2 - 2gy)/u_0^2$$

it is readily found with the aid of (6b) and (8) that

$$E = (Ku_0^4/g) \int_0^{r^2} z^{-\frac{1}{2}}(1 + z)^{3/2} dz \tag{10}$$

where $r = v_0/u_0$. This integral can be converted to an incomplete Beta function by putting $z = w/(1-w)$. It may also be evaluated accurately by expanding the integrand in powers of z, leading to the approximation

$$E = (Ku_0^4/g)(2r + r^3) \tag{11}$$

The error in this curtailed form of the series (whose terms alternate in sign after the second) is less than the magnitude $0.15r^5$ of the third term. In a typical case this error is found to be about 0.3% of the value of E.

With the aid of (6c), equation (11) yields

$$E = K(u_0^2 + \tfrac{1}{2}v_0^2)R_0 \tag{12}$$

The initial kinetic energy per unit mass, namely

$$T_0 = \tfrac{1}{2}v_0^2 = \tfrac{1}{2}(u_0^2 + v_0^2) \tag{13}$$

is finally reduced by air resistance to an approximate value

$$T_1 = T_0 - E \tag{14}$$

   The Effect of Air Resistance on the Length of Jump. Let the range R of the jump be defined in terms of the coordinates (x,y) of G as

   R = the value of x for which y = 0,   t > 0

The range is slightly less than the measured length of a jump, chiefly because (i) the athlete's feet are forward of G when they strike the ground, (ii) contact with the ground does not occur until $y < 0$, which is at a later time than that for which R is calculated. Our object is to compare jump lengths for different air densities; if this is done by comparing R values instead, the error incurred is clearly very small.

   The range R_0 in vacuo of a projectile with initial energy T_0 as in equation (13) is given by

$$R_0 = 2u_0v_0/g \tag{6c}$$

Its range R in air will be calculated as if it were launched in vacuo at the same angle of elevation but with initial energy T_1 as in equation (14). In place of (6c) we will then have

$$R = 2u_1v_1/g \tag{15}$$

where

$$u_1/u_0 = v_1/v_0 = (T_1/T_0)^{\frac{1}{2}} \tag{16}$$

From (6c), (14), (15), and (16) it is readily seen that

$$R = [1 - (E/T_0)]R_0$$

With the aid of (8), (12), and (13) we can write this as

$$R = [1 - K\{1 + (u_0/V_0)^2\}R_0]R_0 \tag{17}$$

The increment in range which accompanies a decrease δK in K is seen from (17) to be

$$\delta R = [1 + (u_0/V_0)^2]R_0^2(-\delta K) \tag{18}$$

Since (10) overestimates the value of E defined in (9), it is easily seen that (18) yields an overestimate of δR.

Beamon's Mexico City Jump. Nonweiler [1956] lists values of 2k (which he calls drag area) for three cyclists at speeds comparable with that of a long-jumping athlete. For his Subject C (whose stature resembles that of Beamon) he lists for the touring and racing positions of the cyclist the respective values

$$k = 0.182 \text{ m}^2, \qquad k = 0.163 \text{ m}^2 \tag{19a,b}$$

Considering the marked difference between the touring and racing postures which are pictured in Nonweiler's paper, the values of k in (19a,b) differ by surprisingly little. Over most of the long-jump flight path an athlete's posture alters by no more than that of a cyclist between touring and racing positions; it is therefore a reasonable approximation to take k as constant throughout the jump. The value in equation (19a) for the touring position with cycle is taken as the appropriate one. Any error incurred will affect similarly the ranges calculated for air of different densities, and so the error in the difference of such ranges will be very small.

Let us consider a jump which in vacuo would have a range

$$R_0 = 8.0 \text{ m} = 26 \text{ ft } 3.0 \text{ in.} \tag{20}$$

This would correspond to a measured jump length of about 9.0 m, or 29 ft 6.3 in.

For use in (13) we will take the estimated value

$$u_0 = 9.45 \text{ m s}^{-1} \tag{21}$$

which corresponds to a sprinting speed of 100 yards in 9.6 seconds. (It may be verified that the final result does not depend critically on the value alloted to u_0, the main conclusion being unaltered if we change u_0 to a value equivalent to 100 yards in 11.0 seconds.) The corresponding values of v_0 and V_0 are found from (6c), (20), (21), and (8) to be

$$v_0 = 4.15 \text{ m s}^{-1}, \qquad V_0 = 10.32 \text{ m s}^{-1} \qquad\qquad (22a,b)$$

The densities of air at sea-level and at the Mexico City altitude of 7400 ft (= 2256 m) are given by Gray [1963, page 3-61] as

$$\rho_1 = 1.225 \text{ kg m}^{-3}, \qquad \rho_2 = 0.984 \text{ kg m}^{-3} \qquad\qquad (23a,b)$$

Taking the mass m of the athlete to be 80 kg it follows from (3), (19a) and (23a,b) that the values of K at sea-level and 7400 ft are respectively

$$K_1 = 2.787 \times 10^{-3} \text{m}^{-1}, \qquad K_2 = 2.239 \times 10^{-3} \text{m}^{-1} \qquad\qquad (24a,b)$$

The change in K accompanying a move from sea-level to Mexico City is therefore

$$\delta K = K_2 - K_1 = -0.548 \times 10^{-3} \text{m}^{-1} \qquad\qquad (25)$$

The corresponding increment in range for the Beamon jump represented by equation (20) is found from (18), (22a,b) and (25) to be

$$\delta R = 0.0645 \text{ m} = 2.5 \text{ in.} \qquad\qquad (26)$$

Discussion. The distance increment of 2.5 inches in (26) is an upper bound on the extra distance which Beamon could have gained from the lower air resistance in Mexico City. While it is much smaller than the gain usually ascribed to this phenomenon, a more accurate solution of (2) by successive approximations (too long to display here) yields a much smaller estimate: 0.7 inches of reduction. Even the total removal of air resistance would add only 4.6 inches to a sea-level jump of record length.

A second effect due to the altitude is the reduction in the gravitational attraction; specifically, the value of g at Mexico City is some 0.07% lower than at sea level. This factor results in an increment of about 0.25 inches in an 8.9 meter jump.

A third effect is the greater forward speed achieved during the run-up due to lowered air resistance. That the resulting increment is negligible can be shown in two ways:

(i) The increase in sprinting speed caused by lowered air resistance is trivial. This statement is supported by the failure of sprinters to lower the world record for the 100 meters [I.A.A.F., 1970] at the Mexico City Olympic Games.

(ii) Most long jumpers do not leave the take-off board at the greatest sprinting speed of which they are capable [Sphicas and Ladany, 1976; Ladany et al, 1975]. They prefer to sacrifice speed (and length of jump) rather than risk the "no-jump" penalty for over-stepping the board or the loss of spring which accompanies a take-off made from behind the board.

The foregoing shows that it is pure myth that Beamon's great jump owed much to high altitude. This conclusion is supported by the failure of all other long jumpers to improve upon their "personal bests" at the Mexico City Games.

With the reduced-air-drag theory discredited, it is natural to seek

other explanations of Beamon's remarkable jump. The writer believes that a
clue is given by the word "Most" in paragraph (ii) above. Beamon was not
one of the conservative majority of dedicated board strikers when he jumped
in Mexico City. He is a sprinter capable of running 100 yards in 9.5 seconds,
and he tried to reach his top speed during his approach. He gave scant
attention to the take-off board and relied solely on his carefully measured
run-up to ensure that he hit the board accurately; his whole effort went
into the jump itself. By a statistical accident his take-off foot landed
in perfect position on the board; the rest of the performance was the in-
evitable consequence of the world's finest long-jumper using a take-off speed
which had never before been achieved. That Beamon himself knew it was a
statistical miracle is evidenced by the fact that he did not even use his
second and third jumps. If his first jump had failed he would presumably
have reverted to a more conventional technique to lessen the risk of missing
the board in his subsequent attempts.

Summary

The air-resistance and gravitational effects of Mexico City's high altitude contributed very little to Beamon's remarkable jump--probably less
than one inch. His dramatic achievement was due to a combination of skill
and luck.

ASSIGNING SWIMMERS TO A RELAY TEAM

by

Robert E. Machol
Based on articles in Operations Research, 1961 and 1970.

The famous physician who discovered a cure for which there was no known
disease has his parallel in many management science discoveries of solutions
for which there is no known problem. Outstanding among these is the assignment problem [Machol, 1957]. Operationally, this problem is to determine the
optimal assignment of workers to tasks, when the performance of each worker
on each task is known. Mathematically, the assignment model can be represented as a linear-programming model in which each coefficient in every constraint equation is zero or unity. It provides a particularly attractive
introduction for students to linear programming because the model is so
simple, and because the Hungarian method [Flood, 1956] provides such elegant
solutions.

Unfortunately there have been no realistic examples of the assignment
problem. The most commonly quoted example, assigning n men to n jobs [Church-
man et al, 1957], depends upon knowing the costs (or performance factors) for
each man on each job, and precise predictions of this sort are simply not
available; if the numbers in the cost matrix are not accurate, one is not
justified in performing a precise cost minimization. The other common example
typically involving routing tractors to pick up trailers [Churchman et al,
1957] is really a degenerate transportation problem, and is unrealistic in
the sense that transportation problems usually involve coefficients other
than unity. The transportation model is of great practical importance; for
if I call Conrail and ask them how much it costs to ship a carload of widgets
from Kokomo to Ishpeming and they say $281.47, that is exactly how much it
is going to cost (because they say so, and the ICC confirms it). But if
our staff psychologist tells me that the performance rating of the third man
on the fifth job is 281.47, I am inclined, on the basis of past experience,
to round it to 300. Finally, numerous problems, such as assigning wire to
terminals or assigning work stations to locations, involve additional compli-
cations which convert then into traveling-salesman or quadratic-assignment
problems which are much more difficult.

The following application appears to fit the model perfectly, and the
appropriate "cost" numbers can be defined and measured precisely. A swimming coach must select from his eight best swimmers a medley relay team of
four, each of whom will swim one of the four strokes (back, breast, butterfly, and free style). The time of each swimmer in each stroke is known (we
here ignore the question of whether expected, minimum, or other times should
be used). The difficulty arises from the fact that each swimmer is good
(i.e. among the best four) in one or more of the strokes, and many of them
in two or more. For example, John Smith can swim 100 yards in 49.3, 61.8,
and 53.7 seconds in the free style, breast, and butterfly respectively; he
does not swim the backstroke. This becomes an 8 x 8 assignment problem. The
rows are the eight swimmers; the first four columns are the strokes and the
last four are dummies. The matrix elements in the first four columns are
the known times and in the last four columns are zeros. For John Smith in
the backstroke, we insert any large number such as 1000 seconds. Clearly,
the sum of "costs" obtained by solving this assignment problem is the minimum relay time, and the best team is determined.

ASSIGNING RUNNERS TO A RELAY TEAM

by

Dennis R. Heffley

Several years ago, Machol [1970] presented an application of the assignment model to a common problem in the world of team sports. In that problem a swim coach must select four of his eight athletes to compete in a four-man medley relay. Each of the eight swimmers has a recorded time in each of the four strokes. The problem is to choose from among the 1680 distinct assignments the one which minimizes total expected time for the event. Defining a_{ki} as the kth swimmer's time in the ith stroke, and letting p_{ki} equal one (zero) if swimmer k is (is not) assigned to the ith stroke, the objective is to assign swimmers to strokes (i.e., choose a set of p_{ki} terms) such that $\sum_{k,i} p_{ki} a_{ki}$ is minimized.

Although the framework employed by Machol (the "linear" assignment model) is useful in situations where one individual's performance in a given role is independent of the way in which other individuals are assigned, there exist many team sports (or other forms of cooperative activity) in which this assumption of independence is untenable -- sports in which the final team performance is not merely the sum of isolated individual efforts. Perhaps it would be useful to look at a very simply example -- an athletic event that is the dry-land counterpart of Machol's medley relay.

Consider the problem faced by a track and field coach who must enter four of his eight sprinters in a quarter-mile relay. While raw speed is an essential ingredient in such an event, smooth and well-timed exchanges of the baton are also important. In general, a given sprinter's time over a 110-yard leg will vary not only with the leg to which he is assigned (since starting, intermediate, and anchor legs require somewhat different talents), but also with the way in which others are assigned. For example, a particular runner's time in the anchor slot might be improved by the assignment of a skilled baton handler of comparable speed to the third leg.

Assume that the kth sprinter's time for the ith leg is determined only by his expected "base" time in that leg (a_{ki}) and by the baton pass from the runner in the preceding leg. This base time might be the sprinter's expected time for the given leg when the preceding runner delivers a flawless handoff. Base time estimates for legs two, three, and four might be obtained by recording the athlete's running-start time in each of these three positions; for the starting leg a block-start clocking is the appropriate base time. The sprinter's total time for a leg will exceed his base time if the runner assigned to the preceding leg provides a less than perfect exchange. If b_{kl} is defined as the amount by which sprinter k's base time in any leg is increased when he receives a handoff from sprinter l, then the problem is to minimize the expression for total relay time,

$$\sum_{k,i} p_{ki} a_{ki} + \sum_{k,l,i,j} p_{ki} c_{ij} p_{lj} b_{kl} \, ,$$

where $c_{ij} = 1$ if the runner in leg i receives a handoff from the runner in
leg j and $c_{ij} = 0$ otherwise (here, only $c_{21} = c_{32} = c_{43} = 1$). Total relay
time for any particular assignment is the summation of relevant base times
and handoff times. Because the p_{ki} variables appear nonlinearly in the above
expression, this particular combinatorial framework has been dubbed the
"quadratic" assignment problem. Note that the linear assignment problem is
a special case of the quadratic assignment problem where all b_{kl} terms and/or
all c_{ij} terms are zero, indicating total independence among participants
and/or roles. For Machol's medley swim relay, all c_{ij} are zero, since no
baton exchange or other direct interaction between team members is involved.

A simple numerical example may illustrate the importance of explicitly
recognizing interdependencies in sports optimizations. Let's first reduce
the dimension of the problem by assuming that only four athletes are available for the event. There are still twenty-four ways in which these runners
can be assigned to the various legs, so it remains to find the best integer
assignment, the permutation which minimizes expected relay time. Suppose
that the array or matrix of base times is given by

$$A = [a_{ki}] = \begin{bmatrix} 10.8 & 10.6 & 10.6 & 10.5 \\ 10.6 & 10.1 & 10.2 & 10.1 \\ 10.5 & 10.1 & 10.1 & 10.2 \\ 10.7 & 10.5 & 10.6 & 10.4 \end{bmatrix},$$

where the kth row gives the base time of runner k in each of the four legs.
Also, let the "compatibility" matrix take the form

$$B = [b_{kl}] = \begin{bmatrix} 0 & .5 & .3 & .5 \\ .3 & 0 & .2 & .6 \\ .3 & .2 & 0 & .7 \\ .2 & .5 & .3 & 0 \end{bmatrix}.$$

The "linkage" matrix that defines the structure of interdependencies in this
event is

$$C = [c_{ij}] = \begin{bmatrix} 0 & 0 & 0 & 0 \\ 1 & 0 & 0 & 0 \\ 0 & 1 & 0 & 0 \\ 0 & 0 & 1 & 0 \end{bmatrix}.$$

Given the above parameters, the expression for total relay time is minimized
when $p_{kk} = 1$, $k = 1,2,3,4$; and $p_{ki} = 0$ for $k \neq i$. This permutation (runner
1 to leg 1, runner 2 to leg 2, etc.) yields an expected time of

$$\sum_k a_{kk} + b_{21} + b_{32} + b_{43} = 42.2 \text{ seconds.}$$

In this particular example, the best assignment is unique, although that
will not always be the case.

Other more intuitive or myopic approaches to the problem may result in inferior assignments. A common "rule of thumb" for positioning runners in such an event has the coach assigning his fastest and second fastest runners to anchor and starting slots, respectively. In this particular example, runners two and three appear to be faster than the other two athletes. From the first column of A we also see that runner three has the fastest base time in the starting leg. Similarly, according to the fourth column of A, runner two has the fastest base time in the anchor slot. Considering the speeds of the remaining two athletes in the two remaining intermediate legs, this "rule of thumb" would probably lead to the assignment in which $p_{13} = p_{24} = p_{31} = p_{42} = 1$. This assignment gives an expected time of 42.8 seconds. Fourteen of the other twenty-three assignments yield expected times that are less than or equal to this figure. In this particular example the optimal assignment calls for placing the two runners with the greatest raw speed in the intermediate legs -- an assignment that would probably be ruled out by any decision rule which ignores interdependencies.

The example demonstrates that the model's explicit recognition of intrateam interdependencies can be beneficial. However, there is another layer of interdependence that the model in its present form ignores. Few team athletic events are run against the clock. The introduction of an opposing team adds a new and interesting dimension to the problem. In addition to considering individual performances in various roles and the compatibility of team members, one must consider another important trait of participants -- competitiveness. Athletes are not uniformly endowed with this characteristic, nor will a given athlete exhibit the same type or degree of competitiveness under all circumstances. Some individuals are capable of exceptional "come from behind" performances, while others develop reputations for being "front-runners." Although the latter is sometimes regarded as a fault rather than a virtue, both are important types of competitiveness. Competitiveness complicates the model by making a runner's time dependent not only upon the way in which he and his teammates are assigned, but also upon the assignment adopted by the opposing team. Formal efforts to account for such factors leads rather naturally into game-theoretic formulations of the problem.

Even if one could successfully incorporate the effects of competitiveness, the model described in this brief paper does not begin to capture all the subtle complexities of a track relay event. Nor does it necessarily imply that the sensitive judgment of an experienced coach can or should be superseded by a mathematical programming algorithm. However, the quadratic assignment model does provide a rather general combinatorial framework which explicitly incorporates some of the interdependencies that characterize team athletics and so many other forms of human activity.

Acknowledgment

The author is grateful to his colleagues, Stephen M. Miller and Roger White, for their helpful comments on a preliminary draft of this paper. Thanks are also owed to Robert Machol for his many thoughtful suggestions.

A THEORY OF COMPETITIVE RUNNING

by

Joseph B. Keller

Based on articles in Physics Today, 1973,
and American Mathematical Monthly, 1974

World records for running provide data of physiological significance. In this article, I shall provide a theory of running that is simple enough to be analyzed and yet allows one to determine certain physiological para- meters from the records. The theory, which is based on Newton's second law and the calculus of variations, also provides an optimum strategy for running a race.

The records give the shortest time T in which a given distance D has been run. Our theory determines this function theoretically in terms of the following physiological quantities: the maximum force a runner can exert, the resistive force opposing the runner, the rate at which energy is supplied by the oxygen metabolism, and the initial amount of energy stored in the runner's body at the start of the race. By fitting the theoretical curve to four observed records, these quantities can be determined and the other records can be predicted. Alternatively, by measuring these quan- tities by independent physiological studies, one can use the theory to pre- dict all the track records to which it applies.

The theory accounts for the main features of the records at distances from 50 meters to 10,000 meters. However, it does not account for the rec- ords at larger distances. These range up to 59 days for 5560 miles, the distance from Istanbul to Calcutta and back, set by M. Ernst (1799-1846), [Lloyd, 1966]. Other physiological factors must be included to account for these records.

Optimal running strategy. A runner's speed varies during a race; we assume that the speed, $v(t)$, is chosen in the way that minimizes the time T required to run the distance D, subject to physical and physiological limitations. We shall determine this optimal speed variation $v(t)$ by for- mulating and solving a mathematical problem in optimal control theory.

The theory predicts that the runner should run at maximum acceleration for all races at distances less than a critical distance D_c = 291 meters. Thus the races at distances less than 291 meters should be classified to- gether as "short sprints" or "dashes." For D greater than 291 meters the theory predicts maximum acceleration for one or two seconds, then constant speed throughout the race until the final one or two seconds and finally a slight slowing down. This result confirms the accepted view that a runner should maintain constant speed to achieve the shortest time, and refines that view by fitting the constant speed to appropriate variable speeds dur- ing the initial and final seconds.

Runners at distances greater than 291 meters often finish with a kick rather than with the negative kick of the optimal solution. This discre- pancy indicates either that they are not doing as well as they could or that the theory is inadequate. Presumably their goal is to beat competi- tors rather than to achieve the shortest time, and that goal influences

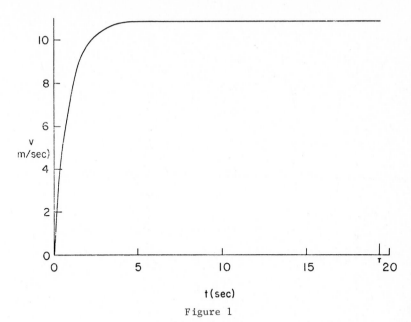

Figure 1

The 220-yard dash. The optimal velocity v(t) is plotted versus t. The propulsive force f(t) = F is maximum throughout the race.

their strategy. But if they ran at the optimal speed determined by the theory, they might do even better at beating competitors. Some trials in which runners attempt to follow the optimal strategy might determine which is the correct explanation, and whether the optimal solution is better than the usual strategy.

In our theory we assume that the resistance to running at speed v is proportional to v. Another assumption, suggested by measurements of R. Margaria and his collaborators [1963a, 1963b], is that the resistance is a constant independent of the velocity. We have worked out the theory in this second case for comparison and found that it leads to a quite unsatisfactory prediction, which indicates that it must be rejected if the other assumptions of the theory are correct.

Formulation of the theory. The length D of a race is related to the time T required to run it by the equation

$$D = \int_0^T v(t)dt \qquad (1)$$

The velocity v(t) is determined by the equation of motion, which we assume to be

$$\frac{dv}{dt} + \frac{v}{\tau} = f(t) \qquad (2)$$

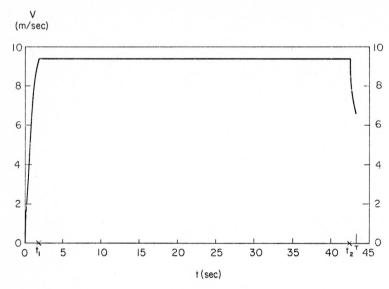

Figure 2

The 400-meter run. The optimal velocity v(t) is plotted versus t. The propulsive force f(t) = F is a maximum during the initial 1.78 seconds. After this initial acceleration, v remains constant until 0.86 seconds before the end of the race when the oxygen supply E becomes zero. Finally E(t) remains zero during the last 0.86 seconds. This can be understood by thinking about a car with a limited amount of gas traveling over a distance D in the shortest time; the fuel should be used up shortly before the end.

In this equation f(t) is the total propulsive force per unit mass exerted by the runner, part of which is used to overcome the internal and external resistive force v/τ per unit mass. It is an assumption that the resistance is a linear function of v and that the damping coefficient τ is a constant. Initially the runner is at rest, so

$$v(0) = 0 \qquad\qquad (3)$$

The force f(t) is under the control of the runner, so we may think of it as the control variable. The runner must adjust it so that T, determined by equation (1), is as small as possible when v(t) is the solution of equations (2) and (3). There are two restrictions on f(t). First, there is a constant maximum force per unit mass F that the runner can exert; so f must satisfy the inequality

$$f(t) \leq F \qquad\qquad (4)$$

Second, the rate fv of doing work per unit mass must equal the rate at which the body supplies energy. This rate is limited by the availability of oxygen for the energy-releasing reactions, which we shall now consider.

Initially there is a certain quantity of available oxygen in the muscles, and more oxygen is provided by the respiratory and circulatory systems. It is convenient to measure the quantity of available oxygen in units of the energy it could release upon reacting. Thus we denote by $E(t)$ the energy equivalent of the available oxygen per unit mass at time t, by E_0 the initial amount, and by the constant σ the energy equivalent of the rate at which oxygen is supplied per unit mass in excess of the non-running metabolism. Then the equations of energy or oxygen balance can be written in the form

$$\frac{dE}{dt} = \sigma - fv \qquad (5)$$

In addition E satisfies the initial condition

$$E(0) = E_0 \qquad (6)$$

Because the energy equivalent of the available oxygen can never be negative, E also must satisfy the inequality

$$E(t) \geq 0 \qquad (7)$$

This is, indirectly, the second restriction of $f(t)$.

Now the runner's problem and ours is to find $v(t)$, $f(t)$ and $E(t)$ satisfying equations (2) through (7) so that T, defined by equation (1), is minimized. The four physiological constants τ, F, σ and E_0 are given, and so is the length of the race D. In other words, the problem is to find the rate of consumption of the initial oxygen supply in order to run the distance D in the shortest time. The solution is not elementary. The relationship between distance D and time T is given [Keller, 1974] by

$$D = F\tau^2[T/\tau + \exp(-T/\tau) - 1], \qquad 0 \leq T \leq T_c$$

$$D = F\tau^2[t_1/\tau + \exp(-t_1/\tau) - 1] + \tau(t_2 - t_1)/\lambda$$

$$+ \tau(\sigma\tau)^{\frac{1}{2}}[-\frac{1}{\lambda}(\tau/\sigma)^{\frac{1}{2}} - \tanh^{-1}\frac{1}{\lambda}(\tau/\sigma)^{\frac{1}{2}}$$

$$- \{1 + [\tau/\lambda^2\sigma - 1]\exp(-(2/\tau)(T - t_2))\}^{\frac{1}{2}}$$

$$+ \tanh^{-1}\{1 + [\tau/\lambda^2\sigma - 1]\exp(-(2/\tau)(T - t_2))\}^{\frac{1}{2}}], \quad T \geq T_c$$

My solution to this problem also gives the result that for D not greater than D_c, where D_c is the critical distance mentioned above, $f(t) = F$, and v increases monotonically. For D greater than D_c, $v(t)$ increases for t less than t_1, v is constant for t between t_1 and t_2, and v decreases for t greater than t_2 until the end of the race, T. Figure 1 shows the optimal velocity $v(t)$ as a function of t for the 220-yard dash $(D < D_c)$. In Figure 2 the optimal $v(t)$ is shown for the 400-meter run $(D > D_c)$.

Table I.

Track Records

Distance D	Time T (record) min:sec	Time T (theory) min:sec	Error (per cent)	Average velocity D/T (theory) m/sec	t_1 sec	$T-T_2$ sec
50 yd	5.1	5.09	-0.2	8.99		
50 m	5.5	5.48	-0.4	9.12		
60 yd	5.9	5.93	0.5	9.26		
60 m	6.5	6.40	-1.5	9.38		
100 yd	9.1	9.29	2.1	9.85		
100 m	9.9	10.07	1.7	9.93		
200 m	19.5	19.25	-1.3	10.39		
220 yd	19.5	19.36	-0.7	10.39		
400 m	44.5	43.27	-2.8	9.24	1.78	.86
440 yd	44.9	43.62	-2.9	9.22	1.77	.86
800 m	1:44.3	1:45.95	1.6	7.55	1.07	1.08
880 yd	1:44.9	1:46.69	1.7	7.54	1.06	1.08
1000 m	2:16.2	2:18.16	1.4	7.24	0.98	1.16
1500 m	3:33.1	3:39.44	3.0	6.84	0.88	1.31
1 mile	3:51.1	3:57.28	2.7	6.78	0.87	1.34
2000 m	4:56.2	5:01.14	1.7	6.64	0.84	1.43
3000 m	7:39.6	7:44.96	1.2	6.45	.80	1.60
2 miles	8:19.8	8:20.82	0.2	6.43	.80	1.63
3 miles	12:50.4	12:44.89	-0.7	6.31	.77	1.80
5000 m	13:16.6	13:13.11	-0.4	6.30	.77	1.82
6 miles	26:47.0	25:57.62	-3.1	6.20	.75	2.10
10000 m	27:39.4	26:54.10	-2.7	6.20	.75	2.12

Comparison of Theory and Observation. Hill [1926] first pointed out
the physiological significance of track records. Since then, many inves-
tigators have tried to extract physiological information from these records
[Lloyd, 1966], but they were hampered by a lack of any theory of running
that correlates the data. Twenty-two world records for distances from 50
yards to 10000 meters are shown in Table I. The first four are taken from
data published by B. B. Lloyd, and the others are from the Reader's Digest
Almanac 1972, page 980. We have determined the two constants τ and F to
yield a least-squares fit of the times given by the theory to the record
times for the first eight races, which we assume to be short sprints. Then
we determined σ and E_0 to give a least-squares fit of the times given by
the theory for the remaining 14 races. In both cases we minimized the sum
of the squares of the relative errors. The values of the four physiological constants obtained in this way appear in Table II. Also shown there
is the value of D_c computed from the theory with these constants. We see
that D_c = 291 meters is between 220 yards and 400 meters; so the first
eight races are short sprints and the other 14 are not. The ratio of the
initial oxygen supply, E_0, to the rate of oxygen supply σ is E_0/σ = 58
seconds. Thus the initial supply is equivalent to the oxygen that would
be supplied by respiration and by circulation in 58 seconds.

Table II

Physiological Constants

$\tau = 0.892$ sec

$F = 12.2$ m/sec^2

$\sigma = 9.93$ calories/kg sec

$E_0 = 575$ calories/kg

$D_c = 291$ m

By using the constants in Table II, we have computed the time given by the theory for each race. The results are shown in Table I, together with the average velocity D/T given by the theory and the values of t_1 and $T - t_2$ for the races with D greater than D_c. The error in time between the theoretical value and the record is also shown in Table I as a percentage. We see that for the short sprints the error is at most 2.1%. However, for the longer races, it reaches 3.1% for the six-mile race. The average velocity given by the theory is plotted against distance in Figure 3, which also shows the average velocities computed from the record times. Note that the initial increase and ultimate decrease of the average velocity is predicted by the theory quite satisfactorily. In comparing the theory with the actual records, it must be borne in mind that the theory uses a single set of physiological constants at all distances, while the record holders at these distances undoubtedly had somewhat different constants from one another. The theory I have presented is too simple, because it omits various important mechanical and physiological effects. It ignores the up-and-down motion of the limbs; it fails to distinguish between internal and external resistance; it does not take into account the depletion of the fuel that uses the least oxygen and the transfer to the use of less efficient fuels; it ignores the accumulation of waste products and the mechanisms of removing them, and it probably ignores some other effects as well. A better theory incorporating some of these effects might be able to account for the records at longer distances, as well as those considered here. In support of such a new theory, measurements of the resistive force and the other physiological parameters are needed.

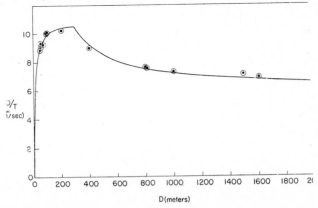

Figure 3

Nevertheless, this theory yields some definite results, and it would be of interest to adapt it to other types of races. Ice skating, swimming, and bicycle races, for example, could be studied, taking into account the special features of each type of race. Another interesting problem would be to determine the influence of hills and valleys on the optimal velocity in longer races; this would require only a modification of the present theory.

A DAY AT THE RACES

by

Paul Bratley

Based on an article in <u>INFOR</u>, 1973

This paper presents some conclusions drawn from the study of 50 evenings of racing, a total of 500 races, at Blue Bonnets Raceway in Montreal during the period between July 22 and September 24, 1972. The track operates six days a week, but there are occasional gaps in our series of results on days when the author was out of town and his wife forgot to buy a newspaper. All these races were run in harness over a distance of one mile.

Wagering at the track in question, as at most North American tracks, is on the pari-mutuel system: the sums wagered on each horse are pooled, the track and the government take a percentage of the pool, and then the remaining amount is distributed to those people who bet on the winning horse.

More formally, suppose that in a race with n horses, an amount a_i, $i = 1,2,...,n$, is bet on each horse. Then the pool will be

$$P = \sum_{i=1}^{n} a_i$$

Suppose the track and the government take a fraction ϵ of the pool. If horse j wins the race, then for every dollar he bets, a winning punter will receive

$$\frac{(1-\epsilon)P}{a_j}$$

dollars in return. From this we may subtract the dollar he paid originally, so that his winnings for a $1 stake are

$$r_j = \frac{(1-\epsilon)P}{a_j} - 1 = \frac{(1-\epsilon) \sum_{i=1}^{n} a_i}{a_j} - 1 \qquad (1)$$

The value r_j, rounded down to the nearest five cents, is called the starting price of horse j. Since this rounding can vary slightly, so can the fraction of the pool taken out before distribution. However, this fraction is always about 18%.

Before each race, as betting progresses, the current values of a_i and r_i for each horse, and of the total pool P, are displayed on the infield board. The display is updated at intervals of thirty seconds or so. By waiting until the last minute it is possible to bet with a very good idea of what the starting price of each horse will be.

The result of a race as reported in a typical daily newspaper gives the starting price r_i for each horse. From (1) above it follows that

179

$$\frac{a_j}{1 - \epsilon} = \frac{\displaystyle\sum_{i=1}^{n} a_i}{r_j + 1}$$

and summing over j we have

$$\frac{1}{1-\epsilon} = \sum_{j=1}^{n} \frac{1}{r_j + 1} \tag{2}$$

This formula enables us to calculate what percentage of the pool was taken by the track and the government.

If the public are able, collectively, to make good estimates of the probability that any given horse will win, then if p_i is the probability that horse i will win, and r_i is the odds on that horse, we would expect

$$p_i(r_i + 1) = \text{constant}, \quad i = 1, 2, \ldots, n \tag{3}$$

The left hand side is the expected return if we bet on horse i. It is sensible to bet on the horse which maximizes this product. But because of the pari-mutuel system, betting on horse i causes a decrease in r_i and hence in $p_i(r_i + 1)$. In equilibrium, therefore, this product will be the same for every horse. Furthermore, the constant in (3) must be $1 - \epsilon$. Thus if the crowd can correctly estimate the probabilities p_i it is immaterial with a pari-mutuel system on which horse one bets: in every case one stands to lose the operator's percentage ϵ.

Suppose however that the crowd estimates the probabilities as being $\hat{p}_i$, $i = 1, 2, \ldots, n$, where in general $\hat{p}_i \neq p_i$. Now equilibrium is reached when

$$\hat{p}_i(r_i + 1) = 1 - \epsilon, \quad i = 1, 2, \ldots, n \tag{4}$$

since the crowd's behavior is obviously determined by its own view of the probabilities. In this case the expected returns $p_i(r_i + 1)$ are not constant, and if for some i, $p_i(r_i + 1) > 1$, then one can bet on that horse in the hope of making a profit.

Since from (2) and (4) we have

$$\hat{p}_i = \frac{1}{r_i + 1} \left(\sum_{j=1}^{n} \frac{1}{r_j + 1} \right)^{-1} \quad i = 1, 2, \ldots, n$$

we may calculate the $\hat{p}_i$ for any race simply from a knowledge of the odds r_i. Now suppose we look at a group of N horses chosen in some suitable fashion. For each of these N horses we calculate $\hat{p}_i$, using data from the appropriate race. Then the expected number of winners among this group, assuming independence, will be

$$\hat{\omega} = \sum_{i=1}^{N} \hat{p}_i,$$ say, while the actual number of winners ω can be counted. A greater or less discrepancy between $\hat{\omega}$ and ω will indicate a greater or less discrepancy between $\hat{p}_i$ and p_i.

Table I was constructed in this fashion, from the starting price and final position of each of the horses in 500 races held at Blue Bonnets over a period of two months. For instance, if the public's estimate of the probabilities is correct, we should have had 47 winners at a starting price of less than 1.00; in fact there were 54.

A glance at this table shows immediately that, while overall the public's error in estimating a horse's chances of winning is quite small, it is nevertheless consistent. In each of the first five groups, that is, for starting prices below 5 to 1, the actual number of winners exceeds the expected number, while for starting prices between 5 to 1 and 25 to 1 the actual number of winners is less than expected. Above 25 to 1 the data is inadequate for a judgment to be made.

We assume that it is possible to bet on a horse knowing what its starting price will be, by waiting until the last minute before betting. Since the public is consistently in error in its estimation of the probabilities, we may hope that a systematic betting strategy will enable us to make a profit.

Table II shows the average profit that would be made by betting $1 on all horses in a particular range of starting prices. This table is based on the sample of 500 races previously mentioned. Although the plethora of minus signs is discouraging to the gambler, certain patterns are discernible, and there are at least some systematic bets which appear to show a worthwhile profit. For instance, backing all horses at odds between 15 to 1 and 20 to 1 in races with 9 or more runners appears to win $0.70 per evening. This can be shown to represent a profit of 27% on the amount actually wagered. Since this average is based on a sample of 130 bets, we may hope that it is not too inaccurate an estimate of the long-term result.

In [Bratley, 1973] an attempt was made to use this kind of information to find a profitable betting strategy, and initial results were encouraging. Sad to say, however, more complete results [Vergin, 1975] lead to a more pessimistic conclusion. It does seem to be true that the public consistently underbets horses with low odds, and consistently overbets horses with high odds. However the resulting bias in the betting is not sufficient to enable an alert gambler to overcome the unfavorable effect of taking out 18% from the pari-mutuel pool to pay the track and the government.

For the moment, therefore, the question of a successful betting strategy remains open.

Acknowledgments. This work was supported in part by Madeline Candor (paid $88.60 for a $2.00 ticket, July 24, 1972), Baby Drummond (paid $85.50, July 31, 1972), Victory Knox (paid $91.50, September 11, 1972), and others.

Table I

Number of winners in 500 races

+ indicates that there were more winners than expected in a group

range of odds	less than 1.00	1.00 to 1.99	2.00 to 2.99	3.00 to 3.99	4.00 to 4.99	5.00 to 5.99
Expected	47.3	107.6	86.9	61.8	42.9	27.4
Actual	54 +	112 +	91 +	66 +	44 +	16 -

range of odds	6.00 to 6.99	7.00 to 7.99	8.00 to 8.99	9.00 to 9.99	10.00 to 14.99	15.00 to 19.99
Expected	21.8	15.9	14.5	11.3	28.3	12.9
Actual	20 -	15 -	14 -	5 -	28 -	12 -

range of odds	20.00 to 24.99	25.00 to 29.99	30.00 to 34.99	35.00 to 39.99	40.00 to 44.99	45.00 or over
Expected	7.0	4.7	2.9	1.7	1.6	3.6
Actual	5 -	9 +	3 +	1 -	5 +	0 -

Table II

Average profit per evening for a $1 stake to win (over 50 evenings racing)

Starting Price

Number of horses in race	less than 1.00	1.00 to 1.99	2.00 to 4.99	5.00 to 9.99	10.00 to 14.99	15.00 to 19.99	20.00 to 29.99	30.00 to 39.99	40.00 or more
5 or less	0.01	-0.09	0.05	-0.50	0.10	-0.18	-0.08	-0.02	-0.04
6	-0.10	0.00	-0.14	-0.51	-0.07	-0.28	-0.24	-0.16	-0.06
7	0.02	-0.43	-0.08	-1.70	-0.51	-0.24	1.31	-0.28	1.15
8	-0.07	-0.19	-0.85	-2.03	0.10	-1.22	-1.25	-0.11	-2.04
9 or more	-0.01	-0.18	-1.95	-1.80	-1.27	0.70	0.47	-0.56	-1.80

by

M. N. Brearley

Early in the 1960 decade the rowing eight of the Australian state of New South Wales surprised spectators by appearing for an interstate race in a racing shell carrying the unusual oar arrangement shown in Fig. 1. In the 1976 Olympic rowing events the Australian boat used the novel oar pattern, but no other nation appeared to have abandoned the inefficient conventional system of Fig. 2.

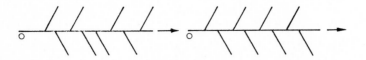

Figure 1   Figure 2

The introduction of the Fig. 1 arrangement was an application of the theory of moments. Anyone who has had a close view of a conventional racing eight in action must have noticed the "fish-tail" behavior of the boat under the oar arrangement of Fig. 2; during each stroke the boat tends to veer first left and then right, the turns being accompanied by noticeable lateral waves from the right and left sides of the rudder in succession. This action increases the resistance to the forward motion, and some loss of speed results. The oar pattern of Fig. 1 does not induce any course deviation, and so provides a more efficient propulsive system.

To make a theoretical comparison of the two oar arrangements, we must first consider the force which an oar exerts on a boat through its point of attachment at the outrigger. During the first half of the "power" part of the stroke the angle θ which an oar makes with the boat is as shown in Fig. 3, with θ increasing to $\pi/2$ as the stroke proceeds. During the second half of the "power" stroke the angle θ exceeds $\pi/2$, as shown in Fig. 4.

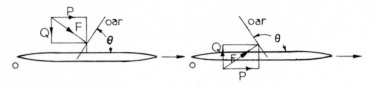

Figure 3   Figure 4

These figures also show the force F which the oar exerts on the outrigger, and its resolution by "parallelogram of forces" into components P and Q which are respectively parallel to and perpendicular to the direction of motion of the boat. The figures depict F as perpendicular to the oar; while this is approximately the case in practice, it is not an essential

feature of the analysis. An important aspect of the force is the reversal
of the direction of the transverse component Q which occurs as θ passes
through the value π/2.

The moments of the forces about any point of the boat, such as the
rudder O, are easily calculated by using the two components P and Q for
each oar. The components P have moments about O which sum to zero over all
eight oars for both Fig. 1 and Fig. 2, since the distance from O to the
line of action of P is the same for each oar. The components Q_r
(r = 1,2,...,8) for the eight oars in the case θ < π/2 are depicted in
Figs. 5 and 6 for the oar arrangements of Figs. 1 and 2 respectively.

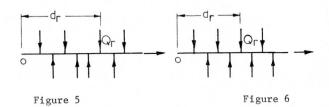

Figure 5   Figure 6

If d_r(r = 1,2,...,8) denote the distances from O to the lines of action
of the forces Q_r, the sum of the moments about O of all the forces is given
by

$$M = \sum_{r=1}^{8} Q_r d_r$$

For the present purpose it is adequate to assume that the oarsmen
exert forces of equal magnitudes, so that

$$Q_r = \pm\, Q, \quad (r = 1,2,...,8)$$

the sign depending on which side of the boat the oar is mounted.

It is easily seen that for the case of Fig. 5

$$M = 0$$

and that for Fig. 6

$$M = 2Q\ell$$

where ℓ is the oar spacing distance and the sign of M has been taken as
positive for definiteness. It is easily verified that these resultant
moments are unchanged if O is taken to be any other point on the centre
line of the boat.

The effect of the moment $2Q\ell$ is to cause the conventionally equipped
boat of Fig. 2 to turn off course (to the left) while θ < π/2. During the
latter part of the power stroke, when θ > π/2, the sign of the moment $2Q\ell$
is reversed, and the boat veers in the opposite direction.

The oar arrangement of Fig. 1 produces no turning moment at any stage
of the stroke, and is therefore more efficient than the traditional one.

MECHANICAL ASPECTS OF ATHLETICS
(Javelin Throwing, Weightlifting, Rowing)

by

Joseph B. Keller

Based on an article in Proceedings of Congress of Applied Mechanics, 1975

Javelin Throwing. In 1953 the world record distance for the javelin throw was 263 feet, 10 inches. In that year a manufacturer of javelins in California, Richard Held, began making them according to a new design. This design involved a redistribution of weight along the length of the javelin, and a flattening of the javelin's cross section, which increased its gliding ability. By 1956 the new javelin was used in the Olympics and the world record had been increased to 281 feet, 2 inches. This increase of over 17 feet in three years is to be contrasted with the increase of less than 20 feet in the next 18 years. The comparison indicates that the remarkable increase within three years was due to the new design.

Weightlifting. Weightlifting was introduced to determine the strongest man. However, it was soon found that the weight L which a man can lift increases as his body weight W increases. Therefore the early weightlifting competitions really determined the heaviest man. This could be done more easily by just weighing the contestants. To eliminate the effect of body weight, lifters are now divided into weight classes, and the strongest man in each class is determined.

These observations suggest the question "How does the lift L increase with body weight W?" To answer it we shall assume that the record holders in all weight classes are equally well developed physically, and that they differ from one another only in size. We shall also assume that the lift L is proportional to the cross-sectional area of a typical muscle. This area is proportional to the square of a typical body length. Since each such length is proportional to the one third power of the body volume, the lift L should be proportional to the two thirds power of the body volume. The volume is proportional to the body weight W, so we conclude that

$$L = KW^{2/3} \tag{1}$$

Here K is a constant.

To test this conclusion, we show in Table I the world record lift L for each weight class W for the two hand press, as of 1969 [Menke, 1969]. In the third column of the table we show the ratio $L/W^{2/3}$. It is clear that this ratio is very nearly constant--much more so than either L or W. The average value of K is 10.83, so that equation (1) with this value of K is the approximate relation between L and W.

By considering the values of $L/W^{2/3}$ we can eliminate the effect of body weight. Then lifters of different weights can be compared. Thus in Table I we see that the record holder in the 148 pound class has the largest ratio, so by this criterion he is the strongest man. In this way weight classes can be done away with, and lifters can compete at any weight.

Table I

World Records For the Two Hand Press (1969)

W	L	$L/W^{2/3}$
123	259	10.48
132	282	10.88
148	317½	11.34
165	327¾	10.88
181	352½	11.02
198	370¾	10.90
Heavy (260)	437½	10.74

$$K = 10.83$$

Their performances can be measured by $L/W^{2/3}$. This idea is not so bizarre. It is a form of handicapping. In yachting, for example, a similar method is used. Different size yachts do compete against one another. A formula based on sail area, hull length, etc. is then used to compare the times they take to traverse a given course.

Rowing. In rowing races, shells with 1, 2, 4, or 8 rowers are usually employed. The more rowers there are in the shell, the faster it goes. The question we wish to consider is "How does the speed v vary with the number n of rowers?" This problem was analyzed by McMahon [1971], whose analysis we shall now present.

The frictional drag force D on a boat is proportional to v^2L^2, where L is a typical dimension of the boat. The volume of the boat, proportional to L^3, must also be proportional to n. Therefore L is proportional to $n^{1/3}$, so $D \sim L^2v^2 \sim n^{2/3}v^2$. The power expended in overcoming the drag is $vD \sim n^{2/3}v^3$. This power is supplied by the rowers, each of whom supplies an amount P. Therefore $n^{2/3}v^3 \sim nP$. On solving this equation for v, we get

$$v \sim n^{1/9}$$

A log-log plot of the time T required for a crew of size n to travel 2,000 meters [McMahon, 1971] demonstrates clearly that a straight line of the form $\log T = \text{constant} -(1/9) \log n$ fits the data quite well, in agreement with the result above.

Editors' Note: The original article [Keller, 1975] also included models of walking and of rope climbing, both continuous and discrete.

SPORT AND THE NEGATIVE BINOMIAL DISTRIBUTION

by

R. Pollard, B. Benjamin and C. Reep

Based on articles in Journal of the American Statistical Association,
Journal of the Royal Statistical Society, and
Revista Brasileira de Estatística

Introduction. In most sports and games, the winner of a particular
contest, be it an individual or a team, is decided both by skill and by
luck. This is readily appreciated by anyone who has played or watched a
sport, but just how big a part each ingredient plays is always difficult to
assess and frequently a topic of much discussion and argument.

There are various ways in which the role played by chance can be investigated. One of these is to examine the frequency distribution of certain events in a game, such as the scoring of a run at baseball. If we are
able to demonstrate that this particular event is governed by the laws of
a probability distribution, then one can say that within the framework of
that distribution, the events are occurring at random. Although the actual
event will occur by chance, the rate at which the event occurs depends also
on the skill of the players involved. The fact that one team may be 'better'
than another, in the sense that it has a higher rate of scoring runs, does
not alter the concept of runs occurring at random, nor rule out the possibility that the inferior team may win due to a random fluctuation.

The negative binomial distribution has been shown to be the governing
principle behind various events in several different sports [Reep and
Benjamin, 1968; Reep et al, 1971; Pollard, 1973 and 1975]. An understanding
of this distribution and of the implication of its widespread applicability
to sport is thus an essential step in elucidating the role chance plays in
determining the winner in any sports contest.

The Negative Binomial Distribution. If we consider a number of individuals, each producing a certain event at the same rate, and if this rate
remains constant, then one would expect the frequency distribution of these
events over fixed time intervals to be Poisson. Suppose, for example, that
each team in a hockey league had probability 1/1800 of scoring a goal in
any one-second interval, and that successive intervals were independent.
Then the distribution of goals scored in a 60-minute game would be Poisson
with mean of 2. Obviously this is never likely to happen in real life.

There are two main ways in which the Poisson model is likely to be
modified. In the first place, the teams will have different rates of scoring goals, varying from game to game. The quality of the opposition, the
present form of the players, and the current league standings are some of
many reasons to suppose that the scoring rate will differ from game to game.
(We assume in this model that the rate of scoring within a game is constant
for each team.) If we consider these different rates as having a general
gamma distribution, then the distribution of goals per game will be negative binomial.

The second modification of the Poisson model also leads to a negative
binomial distribution. If we assume that each team does have the same rate

of goal scoring at the start of each game, but that the rate changes, depending on the occurrence of previous goals in that game, then the distribution of goals will also be negative binomial. In this model, each successive goal scored might increase (or decrease) the probability of a
further goal.

Feller [1943] refers to these two situations respectively as "apparent
contagion" and "true contagion". An excellent account of the negative binomial and other types of contagion is given in Gurland [1959]. Returning
to sport, it seems reasonable to suppose that in the goal-scoring example
there will be both apparent contagion and true contagion. That is to say,
each team will have a rate of scoring goals which varies from game to game,
and throughout each game, depending on previous scoring. We do not have
the exact conditions to expect a negative binomial, but we do have good
reason to suppose that such a distribution will provide an approximate fit
to the data.

For the negative binomial distribution, the probability of r events
over a fixed time interval is given by:

$$P(r) = \binom{k+r-1}{k-1} \; p^k q^r \qquad\qquad r = 0,1,2,3,\ldots$$

where $k > 0$ and $0 < p < 1$ and $q = 1-p$

There are various ways in which the parameters can be estimated (Gurland, 1959). In this paper the simple method of moments is used, giving:

$$p = m/s^2 \quad\text{and}\quad k = m^2/(s^2-m)$$

where m is the sample mean and s^2 the sample variance.

A third way in which the negative binomial may arise is from Bernoulli
trials in which the number of trials is a random variable. For example, in
baseball, if every batter had probability p of making a hit and probability
$q = 1-p$ of making an out, and if there were no double plays or base-running
outs, then the number of hits per half-inning (i.e., before three outs)
would be negative binomial distributed with k=3.

The main observable property of the negative binomial distribution is
that the variance must be greater than the mean, in contrast to the Poisson
distribution for which they are equal. The distribution can be fitted to
observed data in the usual way and a χ^2 test used to measure the fit. The
degrees of freedom can be obtained by subtracting three (one plus the number of parameters) from the number of groups, combining groups if necessary.

Soccer. The first application of the negative binomial distribution
to sport was discovered for soccer, or association football as it is better
known in England [Reep and Benjamin, 1968]. In soccer, the ball is passed
from player to player until a particular player either loses possession of
the ball or makes a shot at goal, or until there is an infringement of the
laws of the game. We can define an 'r-pass movement' as one in which a
player of one side, having gained possession of the ball, starts a series
of r successful passes among members of his own team, after which the rth
recipient loses possession or shoots at goal. If it is assumed that the

probability of pass-success varies from movement to movement, and also
throughout the movement, then we have the situation described above and we
have reason to suspect that a negative binomial might describe the observed
distribution of pass-moves. Table I [Reep et al, 1971] confirms that a
reasonable fit is obtained. Thus the fundamental principle of soccer is
likely to be governed by the negative binomial and hence subject to the
laws of chance within the framework of the distribution.

Though the game of soccer is concerned with pass-moves, the achievement of a long pass-movement is not necessarily the best tactic; in fact
the reverse is more likely the case. Ultimate success is determined by
being able to score more goals than the opposition and it might be that
skill rather than chance is the dominant factor in goal-scoring. Certainly
some teams have higher scoring rates than others, but Table II gives evidence that the occurrence of goals closely follows the negative binomial.
The examples given are for the English Football League in which a large
number of games are played among teams of comparable standard [Reep et al,
1971], and for the World Cup Finals [Pollard, 1975] with only a handful of
games involving teams of more variable attainment.

In top-class soccer nowadays, the average number of goals per game is
less than three, so that matches are usually decided by very narrow margins,
typically 1-0 or 2-1. Thus a game between teams of not greatly differing
scoring rates is likely to be decided by chance. Only over a long series
of matches, such as a league lasting the whole season, will the 'best' team
be able to overcome the effect of random fluctuations and emerge as deser-
ving winners. The soccer world, players, coaches, press, and fans alike
refuse to accept this fact, and continue to give overwhelming significance
to the results of individual games--all this despite such clearly fortuitous
results as the United States' defeat of England by 1-0 in the 1950 World
Cup in Belo Horizonte, Brazil.

Football. Touchdowns in football are analagous to goals in soccer.
Although other factors go into making up the total score, Pollard [1973]
has shown that the distribution of collegiate football scores, as presented
by Mosteller [1970], is roughly negative binomial (Table III). Touchdowns
are about twice as frequent as goals at soccer; even so, chance must play
a large part in determining the winners of individual games.

Baseball. Runs per half-inning, as recorded from World Series games
over 14 years, produce a close fit to the negative binomial, as shown in
Table IV [Reep et al, 1971]. The model described earlier seems to be applicable in this case, though Lindsey [1961] did suggest a more elaborate model
to describe runs per half-inning and managed to obtain a good fit to his
data, although scarcely better than that given by a simple negative binomial
distribution. Nearly three quarters of the observed data fall into the
0-run category, making a run a rare event.

Hockey. The distribution of goals per match by individual teams in
the National Hockey League has been shown (Table V) to be closely negative
binomial [Reep et al, 1971]. This is not surprising, even though goals are
more frequent than at soccer.

Tennis. The analogy to the soccer pass-movement in tennis is the length
of a rally. However, since the server is more likely to win a point than
the receiver, and since this is more likely still if his first serve is in,

four types of rally can be distinguished: first serve in, won by server; first serve in, lost by server; second serve in, won by server; second serve in, lost by server. Data have been recorded between 1967 and 1973 in the men's singles finals and semifinals matches played at Wimbledon. However none of the four types of rally produced a close fit to the negative binomial (Table VI).

This is perhaps due to the fact that the advantage of serving disappears after the first few shots have been played. In both examples given, the number of points won at 7 strokes is many fewer than expected; a similar phenomenon occurs for the points lost by the server, making the distributions far from smooth.

Cricket. The distribution of individual batsmen's scores is discussed in 'Cricket and Statistics' (p. 129) and reasons are given there why a negative binomial model might not be appropriate. However, these reasons are less valid if we consider not individual batsmen's scores, but partnerships; that is, the combined scores of pairs of batsmen while batting together. The surfeit of high individual observed scores scarcely affect the distribution of partnerships, since other batsmen are likely to be dismissed while the big score is being made. In fact, the distributions of partnership scores produce much closer fits to the negative binomial than do scores of individual batsmen. Table VII gives the distribution of all partnerships for a team in the English County Championship throughout a season and involves the performances of over 20 players.

Conclusion. The negative binomial distribution has been shown to apply to data from a variety of different sports, the closeness of the fit depending on a number of factors. It seems that the occurrence of infrequent events involving groups of players or teams is most likely to give rise to this distribution; scoring of goals in a soccer match or runs in a baseball half-inning are examples. Data taken from performances of individual players do not give close fits to the negative binomial, and in these cases the individual skill of the players may be more significant than chance in determining occurrences of the event. Batsmen's scores at cricket and the length of a rally at tennis provide examples. It is interesting to note that as soon as the number of players involved is increased, as was the case of partnerships in cricket, the fit becomes much closer.

With the negative binomial distribution underlying so many events in sport, it seems likely that the role played by chance in determining the outcomes of these events has been underestimated. In particular, a game such as soccer, the result of which is dependent on occurrences of comparatively rare events (goals), is especially likely to be governed by chance.

Table I

SOCCER. Passing Movements.
42 English First Division Football League Matches. 1957-58.

Number of passes (r)	Observed	Expected
0	10,187	10,143
1	6,923	7,022
2	3,611	3,553
3	1,592	1,578
4	608	651
5	280	257
6	107	98
7	33	37
8	9	13
9 +	11	8
Total	23,361	23,360

Mean	1.02
Variance	1.50
$P(\chi^2)$	.13

Table II

SOCCER. Number of Goals Scored Per Game by Individual Teams.

Number of goals (r)	English First Division Football League 1967-68		World Cup 1974	
	Observed	Expected	Observed	Expected
0	225	226.6	28	31.4
1	293	296.4	24	20.0
2	224	213.9	13	11.4
3	114	112.6	5	6.2
4	41	48.3	4	3.3
5	15	17.9	0	1.8
6	9	5.9	0	0.9
7 +	3	2.5	2	1.0
Total	924	924.1	76	76.0
Mean	1.51		1.28	
Variance	1.75		2.54	
$P(\chi^2)$	.57		.42	

Table III

FOOTBALL. Number of Points Scored Per Game by Individual Teams.

U.S. Collegiate Football 1968.

Number of points	r	Observed	Expected
0 - 5	0	272	278.7
6 - 11	1	485	490.2
12 - 17	2	537	509.1
18 - 24	3	407	406.6
25 - 31	4	258	275.9
32 - 38	5	157	167.3
39 - 45	6	101	93.5
46 - 52	7	57	49.0
53 - 59	8	23	24.4
60 - 66	9	8	11.7
67 - 73	10	5	5.4
74 & over	11 +	6	4.3
Total		2316	2316.1

Mean	2.58
Variance	3.79
$P(\chi^2)$	.59

Table IV

BASEBALL. Number of Runs Scored Per Half-Inning

World Series. 1947-1960

Number of runs (r)	Observed	Expected
0	1023	1017.7
1	222	227.5
2	87	86.4
3	32	37.3
4	18	17.1
5	11	8.1
6	6	3.9
7 +	3	3.9
Total	1402	1401.9

Mean	0.48
Variance	1.04
$P(\chi^2)$	.70

Table V

HOCKEY. Number of Goals Scored Per Game by Individual Teams.

National Hockey League. 1966-67

Number of goals (r)	Observed	Expected
0	29	27.6
1	71	68.8
2	82	91.4
3	89	85.9
4	65	64.1
5	45	40.4
6	24	22.3
7	7	11.1
8	4	5.0
9	1	2.1
10 +	3	1.3
Total	420	420.0

Mean	2.98
Variance	3.57
$P(\chi^2)$	.75

Table VI

TENNIS. Number of Strokes Per Rally

Wimbledon Men's Singles Matches. 1967-73.

Number of strokes	r	First serve in. Point won by server.		Second serve in. Point won by server.	
		Observed	Expected	Observed	Expected
1	0	591	592.1	219	226.4
3	1	389	399.2	234	228.0
5	2	212	182.8	130	116.0
7	3	53	70.6	29	39.7
9	4	20	24.7	6	10.3
11	5	11	8.1	3	2.2
13 +	6 +	5	3.6	2	0.4
Total		1281	1281.1	623	623.0

Mean	0.89	1.02
Variance	1.17	1.03
$P(\chi^2)$	.01	.07

Table VII

CRICKET. Number of Runs Scored Per Partnership

Middlesex in English County Championship. 1956

Number of runs	r	Observed	Expected
0 - 19	0	269	265.3
20 - 39	1	71	72.7
40 - 59	2	24	30.2
60 - 79	3	15	14.0
80 - 99	4	9	6.8
100 - 119	5	5	3.4
120 & over	6 +	3	3.6
Total		396	396.0

Mean	0.62
Variance	1.40
$P(\chi^2)$	.51

COMPUTERS AND SPORTS· FROM FOOTBALL PLAY
ANALYSIS TO THE OLYMPICS GAMES

by

J. Gerry Purdy

Based on articles in <u>Datamation</u>, 1971, <u>Research Quarterly</u>, 1974, and others.

Have you ever wondered if it would be possible to develop mathematical theory and associated computer programs which could pick the winning
play in the Superbowl? Although it is conceivable that such theory and
programs might be developed, a computer system, no matter how sophisticated,
can only provide information for the coaches, participants, and spectators.
In the final analysis, the athletes themselves must actually perform the
sporting event.

This chapter describes computer systems that have been employed to
provide interesting--and often useful--information for many different sporting activities. These systems have varied from computer-controlled stadium
scoreboards which display statistics, animated cartoons, and instant re-
plays to systems which perform rather intelligent functions such as predicting the outcome of a sporting event or telling runners how they should
train.

Football Play Analysis. The prominence of football in the United
States has provided motivation for the development of computer systems to
aid the coaching staff in making analysis of the game plays. Many professional and college football teams utilize computer programs to accumulate tendency statistics about plays conducted during a game. The coaching staff of a football team will analyze the plays of both the offense
and defense of their opponents as well as the plays of their own team.
This procedure allows the coaching staff to become aware of the trends
which the offense and defense make within given situations. Computer programs analyzing the offensive plays have been used more than the analysis
of the defensive formations. The offense is less likely to change from
week to week, whereas the defense is often adjusted from game to game based
on the type of offense used by the forthcoming opponent. However, the
gathering of tendency statistics about defensive activities such as blitzes
(a charging defensive back) and stunts (unusual movements by the defensive
line) can provide useful information for the offense coach.

Football play analysis usually proceeds as follows. After a football
game, the film crew creates two films· one film is made with all the
offense plays for one team (and the defense plays for the other) and another
film is made with the offense plays of the second team (and the defense
plays of the first team). These films are usually available by the morning
following the game. The assistant coaching staff views these films and
"breaks down" each individual play into its constituent parts. These parts
include the play number, down, distance (to go for a first down), field
position, hash-mark, formation, motion, play type, hole, pass zone, gain,
consistency, and situation. A sample sequence of typical football plays
is shown in Figure 1.

From the input play information, the computer program sorts and accu-
mulates various statistics. An example of the output of a computer foot-

INPUT PLAYS
=========

PLAY NO.	DWN	YARD TO GO	YARD LINE	HASH	FORM.	PLAY	MTN.	HOLE	GAIN	PLAY TYPE	CARRIER	RECEIVER	PATTERN	ZONE	PASS RESULT	SCORE	ERROR	CONS	SHIFT	QUAR	REL. SCOPE	SITU	AUTO
1	1	10	-28	M	GRMR	16TRP	6	6	6	X	PR						+		1	T		N
2	2	3	-33	M	GRNP	72	2	3	2	P	QB	FL	OT	SF	INC		+		1	T		N
3	3	5	-33	M	GRMR	13		8	20	R	FB						+		1	T		SH
4	1	10	-35	R	GRMF	18RLS	3	1	4	R	FB						+		1	T		N
5	2	4	45	L	BLUR	24ISO	4	4	3	R	FA						+		1	T		N
6	3	1	41	L	PLUP	28ISO	1	1	1	R	FA						+		1	T		SH
7	1	3	38	P	PNT					K	P3						−		1	T		N
8	1	10	37	L	R			8	9	R	FB								1	T		N
9	2	1	-26	R	GRMR	18BLS	4	4	5	R	FB	SE	OT	W3	INC		−		1	T		SH
10	1	10	-35	R	RLUP	24ISO	3	3	0	R	QB	SF	CU	WC	INC		+		1	T		N
11	2	10	-40	R	GRMR	73		5	5	P	QB	SE	CU	WC	COM		+		1	T		N
12	3	10	-40	L	GFMR	16PA	5	18	18	P	QB	SE	OT	W3	INC		+		1	T		N
13	1	10	39	L	GRMR	16PA	8	0	0	P	QB						−		1	T		N
14	2	10	39	L	GRMR	73	3	8	18	P	FB						+		1	T		N
15	1	10	18	R	GPMR	18RLS	2	2	2	P	FA						+		1	T		N
16	2	8	16	R	GPMP	93	3	4	6	P	QP	SE	OT	W3	COM	TO	+		1	T		GL
17	1			R	GRMR			7								XP			1	T		GL
18	3	1	14	M	RLUL	27ISO	3	1	3	R	QP								1	T		GL
19	2	6	3	L	RLUP	27ISO	1	2	2	R	PR						+		1	T		GL
20	3	1	1	M	BLUP	24ISO	1	0	1	R	PP						+		1	T		GL
21	1	1	1	T	BLUP	26DPT		0	0	R	QB						+		1	T		GL
22	4	0		R	PAT					K									1	A		N
23	1	10	-28	R	GPNL	73		3	−5	R	QR	SF	OT	W3	SAC		−		1	A		N
24	2	15	-23	M	BLUL	22NPW	3	2	−7	R	RR						+		1	A		N
25	3	8	-30	L	RLUL	78		8	44	R	QB	FL	ST	S3	COM		+		1	A		SH
26	1	10	76	R	GRNP	73	2	8	7	P	QB	SE	OT	W3	COM		−		1	A		SH
27	2	3	19	L	GRMR	18BLS	3	8	0	R	FB						+		1	A		N
28	3	3	19	M	RLUP	21ISO	4	0	−2	R	RB				FG		+		1	A		N
29	1		21	M	FG			1	0	K	QB						−		1	A		N
30	1	10	16	R	GRMR	41SWP	1	1	15	P	RR	SE	PO	W3	COM	TO	+		1	A		GL
31	1	3	15	R	GRMF	16PA	3	7	15	P	QB	SF	PO		COM	TO	+		1	A		GL
32		9	10	P	GRMF	16PA	1	7	0	P	QB					XP	+		1	A		GL
33	4			M	PAT			7	3	K	RR						+		2	A		N
34	1	10	-33	R	RLKL	47	U	3	22	R	FA	SE	OT	W3	COM		+		2	A		SH
35	1	7	-36	R	GRML	19SWP		8	22	R	QB	SF					+		2	A		N
36	3	5	35	M	GRML	73		8	13	R	FB						+		2	A		N
37	1	10	22	R	RLKR	47	O	7	8	R	RB						+		2	A		N
38	2	4	16	R	BLUL	27ISO	T	7	7	R	RB						+		2	A		N
39	1	4	1	R	BLUR	27ISO	T	7	7	R	RB	SE	OT	W3	COM		+		2	A		N
40	1		1	R	GRML	41SWP		0	−2	P	RB						−		2	A		N
41	4	3	-3	M	FG			7	0	K					FG		+		2	A		N
42	1	10	-33	M	BLKL	0 47		9	−2	X	RB						+		2	A		GL
43	2	8	-20	L	GRNF	19SMP		4	1	R	FB	SE	ST	W3	INC		♦		2	A		GL
44	3	23	-23	L	GRNR	27DRW		0	0	P	PB						−		2	A		GL
45	4	31	-12	P	PNT	79		0	0	K					SAC		+		2	A		N

Figure 1

Sample input to a computer football play analysis program. YARD LINE is measured from the 50 (negative means the ball is on their own 0 to 50 yard line). FORMATION and PLAY are codes employed to describe the alignment of the offensive personnel and the actual offensive play respectively. MTN. - player motion, i.e., movement in the back-field before the ball is in play; HOLE - a number indicating the position on the line through which the ball carrier ran; GAIN - positive or negative yards gained in the play; PLAY TYPE - run, pass, kick or penalty.

```
************************************************************************
                      LIST OF PASSES BY ZONE
         GIVING FORMATION,PLAY,RESULT,RECEIVER,PATTERN,GAIN
                   FORMATION DIRECTION - RIGHT
************************************************************************
   5*                      6*                        6*           7*
*GRNR  -73    INC SE OT    0 *  GRNR -16PA  COM SE PO 15  *BLUR T-26BTL RUN FL PO  7 *
*GRNR  -73    INC SE OT    0 *  GRNR -16PA  COM SE PO 15  *GRNR -92     COM FL OT  0 *
*GRNR  -93    COM SE OT    6 *  GRNR -76    INC FL PO  0                 *          0 *
*GRNR  -73    COM SE ST  -11 *  GRNR -75    INT FL PO  0                 *          0 *
*GRNR  -79    SAC SE ST  -10 *  GRNR -73DLY COM TE CR  7                 *          0 *
*GRNR  -79    SAC SE ST      *  BLUR -59ROL COM TE CR 18                 *          0 *
*GRNR  -59ROL COM SE OT    9 *                                          *          0 *
*GRNR  -73    COM SE OT   10 *                                         6*          7*
   5*                      6*
************************************************************************
   1*                      2*                        3*           4*
*GRNR -16PA -INC-SE-CU-  0 * *GRNR -16PA -INC-SE-CU-  0 *  *GRNR -72 -INC-FL-OT-  0 *
*GRNP -16PA -COM-SE-CU- 18 * *GRNP -16PA -COM-SE-CU- 18 *       -72 -INC-FL-OT-  0 *
   1*                      2*                        3*           4*
************************************************************************
```

Figure 2

Sample output of a computer football play analysis program. This report shows the actual areas in which the passes were thrown. Zones 1-4 represent short passes; zones 5-7 represent longer passes. Even though the formation was strong to the right (tight end was on the right side of the line), the quarter-back had a tendency to throw long passes to the left side; the defensive secondary possibly could expect this tendency to occur in future games.

ball play-analysis program is shown in Figure 2. Typically, from 20 to 100 different reports are generated by most football play-analysis computer programs. The coaching staff takes the computer-generated reports and analyzes them for tendencies. These tendencies are summarized and included in the coaching-staff scouting report for the next week; this report usually is available one or two days following the game.

Commercially available football play-analysis systems were utilized as early as 1966. William Witzel developed one of the first systems for the Washington Redskins [1968] and implemented his system for many other pro teams including the Chicago Bears, Dallas Cowboys, San Francisco 49ers, and Atlanta Falcons. Most of the NFL teams now use computer programs to analyze plays. These programs, usually programmed in either FORTRAN or COBOL, have been modified during the ensuing years to incorporate changes to satisfy the coaching staff. Joe Gardino's Los Angeles based APEX organization has a number of pro team clients.

Dr. Frank Ryan, well known quarterback for the Cleveland Browns in the 1960's and now Director of Information Systems for the House of Representatives, developed one of the most sophisticated football play-analysis systems. His system, called PROBE, was a general file-definition and report-generator facility, coded in ALGOL 60, and operated on a Univac 1108. This system produced various plots, histograms, and sorts as requested by the report generator (see Chapter by Ryan, Francia, and Strawser).

Numerous attempts to market football play-analysis programs at the college and high-school level also have been made. These include Sam Huff's Computerized Scouting System, Ed Flaherty's Scout-Aide System [1972], and COMPU-SCOUT, which utilizes an IBM Porta-Punch for their clients in Texas and northern California. Some Universities such as Dartmouth [Blackman, 1970] and Tennessee have developed their own play analysis systems. Quarterback Virgil Carter examined successful plays resulting from a given situation [Carter and Machol, 1971; Smith Corp., 1973]. Casti [1971] made an attempt to develop a theoretical dynamic-programming model for optimal football play selection, although the model was not evaluated.

Others have examined the possibility of predicting the outcome of football games based on the statistics recorded during past performances. Bud Goode of Hollywood, CA, has accumulated statistics not only on football games, but also on other sports including basketball and hockey. He has used various regression procedures to predict which team will win an event. His analysis for the 1974 Super Bowl picked the Miami Dolphins by 9 points over the Minnesota Vikings [Marshall, 1974], the pre-game favorite. The final score was Miami 24 - Minnesota 7.

Football Player-Selection Systems. Computer systems have also been utilized to aid NFL coaches in selecting the eligible college players during the annual draft [Furlong, 1971; Libby, 1971; Rathet, 1971; Sayre, 1968; TV Guide, 1971]. The purpose of these systems is to provide a consistent and reliable rating of the available athletes. A number of scouting combines have been formed to allow more efficient use of scouting personnel. Each of these combines has one or more representative scouts visit almost every college in the country that has a potential pro football player. Each scout generates a report on each player examined. The scouting reports are merged together, and the data is stored in a computer file. The player selection program ranks the players by position as well as over-all athletic ability. Finally, the coaching staff refer to these ratings during the draft.

Presently, there are three scouting combines: TROIKA (49ers, Rams, and Cowboys), BLESTO (Bears, Lions, Eagles, Steers, Vikings, Colts, Bills, and Dolphins) and CEPO (Browns, Cardinals, Packers, Redskins, Giants, Falcons, Saints, and Patriots). The other teams perform their own scouting individually, although the Chargers have utilized some of the scouting personnel of the TROIKA organization.

The scouting process costs the NFL clubs over $3 million every year [TV Guide, 1971]. For this money, more than 12,000 football players are evaluated and rated each year. The combines serve many teams and are intended to be more cost-effective than each team doing the whole job by itself. According to Jack Butler of BLESTO, "One team could never afford to scout the whole country. We scout every school and perform detailed analysis - more than any one team could by itself, and it's cheaper for each team."

The computer processing of the TROIKA organization is performed by Optimum Systems Corporation (OSI) of Santa Clara, CA. The TROIKA scouts complete their reports using mark-sense scanner sheets. OSI converts them into punched cards using an IBM 1232 Optical Mark Page Reader - 534 Keypunch System. The cards are then input to their player-selection program operating on the OSI IBM 370 computer, and the generated reports are given to the participating clubs. Each club's report is different because each club provides additional scouting reports completed by their own staff.

How good are the player-selection systems? Generally, the player-selection systems have been successful in providing consistent and unbiased ratings of the available college players. In 1968, the rating report for the Dallas Cowboys showed Calvin Hill to be the best athlete available when it came time for Dallas to choose. Even though he was an unheard-of player from Yale, Dallas chose him, and he went on to become a star in the NFL. Sometimes the ratings mislead the coaches, but this usually only happens when the player has an injury or decides not to perform to his full potential. It is important to realize that these systems do not provide substantially more information than available from the scouting reports. What the rating reports do provide is a means to rank the athletes in a consistent manner with most of the bias eliminated (each scout has a weighting factor specified by the coach of the team).

Track and Field. Computers have been utilized in many different ways in track and field. Purdy has utilized a computer to generate a decathlon scoring table [1972, 1974a, 1974c, 1975a, 1976a, 1976b]. In the decathlon, athletes compete in 10 different events, and the winner is determined from the accumulated value of his performances. A scoring table is a mechanism to compare the quality of performance both within and between track-and-field events. The measure of the quality of a performance is referred to as a point score which is an abstract numeric value. A scoring table is a mapping from a performance mark into a point scale. For historical reasons, performance marks usually have ranged from zero to 1000 points, with the zero point level equivalent to a very poor performance, and the 1000 point level at or near the performance mark achieved in international competition.

The International Amateur Athletic Federation (IAAF) has officially adopted four scoring tables for men's events: 1912, 1934, 1952, and 1962 [IAAF, 1962]. A modification to the 1962 tables took into consideration photoelectronic timing to 1/100 sec for the 100 m dash and 110 m high hurdles [IAAF, 1971b]. The two adopted scoring tables for women were in 1954 and 1971 [IAAF, 1971a].

This author developed a mathematical model [1975a] which relates performance marks to point scores; the derived equation has a linear and an exponential term. The performance marks are expressed as a velocity for running and
hurdle events, height for vertical-jumping events, and distance for other
field events. This model has been evaluated [1976a] and analyzed [1976b] to
show that the current IAAF scoring table should be replaced by a newer table
with a more equitable relation between performances at the different point
levels.

Scoring tables have to be modified periodically to reflect the changes
in relative athletic performance. Perhaps the apparently large improvements
in the distance running events over recent years will be small compared to
the improvements made during the next 50 to 100 years. A recent study by
Ryder, et.al. [1976] projected that in 2028 A.D., the world record for the
mile would be 3:30 and the 26.2 mile marathon would be 1:53:13, or an
average of 4:18 per mile. Only the long-jump record of 8.90 m may not be
equalled or surpassed (see Brearley's article).

Gardner and Purdy have developed a model for running interval training
[1970a, 1970b]. A runner learns from track publications about the specific
interval workouts that the world class athletes are doing, but he has no
guidelines to tell him how he should do such workouts at his own level of
ability. In addition, many track coaches desire to provide each runner
with a personalized interval workout. To obtain an interval workout, the
athlete establishes his level of ability from a scoring table. Then, the
athlete refers to the "pacing table" assigned for his point level. These
pacing tables contain calculated submaximum times for various interval distances. The suggested number of repetitions and amount of rest between repetitions are also listed. From this information, the athlete is given all
the necessary information for a personalized interval workout.

Recently, a computer program was designed to quantify the energy cost
for endurance exercise [Purdy, 1975b]. Dr. Kenneth H. Cooper, in his popular books concerning aerobics [1968, 1970, 1973], has developed a system of
endurance exercises based on a health-maintenance program called aerobics.
Aerobic exercises stimulate the heart, lungs, and cardiovascular system.
Aerobics is a system of graded aerobic exercises designed to elicit an improvement in the cardiovascular system. A scale was devised which assigns
points in relation to the benefit of the exercise performed. At the Aerobics Center in Dallas, a computer program is utilized to calculate the
aerobics points from individual exercise logs. By this technique, the computer has been used to calculate aerobic points for over 1000 people on a
regular basis. This system provides the ability to document the exercise
habits of a large number of people over a long period of time. It is hoped
that such data can be used to help determine the relationship between endurance exercise and overall cardiovascular health.

Osborne and Jones [1971] of Endicott, NY, developed a timing and recording system for use during cross-country races. As each competitor crosses
the finish line, an interrupt signal is sent via a time-sharing terminal to
a computer which records the runner's time. The competitor numbers are entered in the order of how they finished, and the program generates a listing of the final results, including team scores to determine the winner.

A Hewlett-Packard 9830 minicomputer has been programmed to score the
decathlon at the competition site [Purdy, 1976c]. The 1962 IAAF scoring

tables were placed on cassette tape, and the program was written using the
BASIC language. The program accepts the performance marks for each competitor in an event, calculates the point score from the performance values,
and ranks the competitors after each event. This program was utilized to
help score the 1974 Texas Relays and NCAA decathlon competitions. It demonstrated that a minicomputer could be easily transported to the athletic
field and maintain accurate records of the decathlon competition.

Information-retrieval systems have been developed to support the Olympic
Games. At Munich in 1972, the Siemens Company implemented an information-
retrieval system which provided Olympic sports information for the press and
spectators. From a database of winning performances by the athletes in all
past Olympic Games and the best performances of the current Olympic competitors, a query facility was developed which allowed the user to make a
request containing a number of logical conditions. For example, it was
possible to request a listing of all past Olympic 1500 m champions who had
winning times less than 3:50. In addition, a computer system directly con-
trolled the display board in the main Olympic stadium in Montreal, similar
to the computer-driven display boards in most large stadiums in the U.S.

Walker & Kirmser of Kansas State University developed a model of the
mechanics of pole vaulting [1973]. Their computer evaluations show that it
is not unreasonable to expect someone to vault over 20 ft. In 1975 Honeywell Corp. provided support for administrative aspects of the Boston Marathon [Computerworld, 1975b].

Baseball. Computers have been employed to analyze, model, and simulate many facets of baseball. In 1970, a computer was used in an attempt
to determine the all-time best baseball team [Rathet, 1970]. A computer
program was developed to simulate the greatest teams playing each other in
an all-time world series. Eldon and Harlon Mills of Computer Research and
Sports of Princeton, NJ, programmed a simulation in which the 1927 Yankees
beat the 1961 Yankees in the final playoff series.

Mills and Mills [1970] developed a computer-based system which calculates the Player Winner Average that they contend is the best available
measure of baseball ability. Others have attempted to use past statistics
with a computer to arrive at the optimal batting order, the type of pitches
to hit, the type of pitch to throw against a batter, and other offense-de-
fense strategies [Chaitin, 1971; Pavlovic, 1971].

The Atlanta Braves have used an on-line statistical reporting system
developed by Honeywell. This program, written in FORTRAN IV for the Honeywell 1648 time-sharing computer, was developed by Gary Williams and Susan
Gerald. Miss Gerald describes their program: "Information is input to
the machine as events occur during the game; game situations have been coded
for ease of input-BB signifying base on balls, 1B a single, etc.--and files
are instantly updated according to player number. At any point in the game
the operator may interrupt the data input to request short statistical
print-outs that can include an up-to-the-minute line-up stat sheet or an
up-to-date sheet on any player, any combination of players, or the entire
team. In addition, the system maintains and will print upon request files
of special situations for pitchers and selected batters. Also, the system
includes a short routine that will respond to any question in a conversa-
tional mode. The use of this system has already pointed out several interesting facts about individual players as well as the team as a whole. For

example, through comparison of the statistics concerning Braves batting
against left-handed pitchers versus batting against right-handed pitchers,
we found that our Latin-American players as a rule hit better against the
right-handed pitchers."

Most baseball fans are familiar with the voting procedure for the
yearly All-Star Game. Marden-Kane, Inc., supplies tabulating cards where
the voter punches out a hole next to the player of his choice. The cards
are then processed by Marden-Kane and the All-Star Game rosters are generated. Finally, SDC in Santa Monica, CA, has offered a SPORT/STAT Service
for use by local softball/baseball clubs to keep track of their statistics
and standings.

Boxing. In 1970 a computer was utilized to simulate a fight between
Rocky Marciano and Muhammad Ali. The simulation was developed by Systems
Programming Services, Inc., in Miami under the promotion of Murry Woroner.
The computer program was written in FORTRAN IV to run on an NCR 315 computer. According to Henry Meyer of SPS, the database for the simulation
was generated from past statistics of each boxer plus subjective evaluations:
"We had boxing experts such as Nat Fleisher, the Dundees, and members of
the World Boxing Historian's Association throughout the world evaluate some
55 characteristics of each fighter on a scale of from 1 to 10. These characteristics included such things as speed and hardness of various types of
blows, ability to deliver and avoid punches, stamina, courage, etc. The
ratings were then statistically reduced to a weight for each factor for each
fighter. Of interest, courage was the most important of all."

In testing the system, a fighter was occasionally pitted against himself; the results were rather interesting. Many times the fighter really
out-classed himself. Often his weakest defense was against his best blow.
Over 70 rounds of simulated boxing were filmed. The computer simulation
was run, and the filmed rounds which most closely matched the computer
simulation were included in the 15-round match that was staged. After each
round, a computer output was flashed on the screen summarizing the score.
Marciano won a 15-round decision, but Ali, of course disagreed.

Basketball. A rating system of performance by players of the New Orleans Jazz has been developed by Dr. Cecil Hallum [Bordelon, 1975]. His
rating system takes the form of percentiles which measure the relative performance in minutes played, field goal and free throw percentages, offensive, defensive, and total rebounds, assists, and total points.

Hallum takes the individual game statistics and extrapolates them
linearly to full-game values. He then calculates a standard score which
compares each player's performance with the other 23 players on the two
teams, and finally he converts the standard scores into percentiles. His
computer program generates a report ranking each player's percentile within
a given category and also ranking the categories for each player.

The conversion from standard scores to percentiles makes an assumption
that the average game statistics follow a distribution that is representative of the NBA. Hallum has run Chi-square and Kolmogorov-Smirnov tests
comparing the distribution of game averages with the entire NBA. These tests
showed that game averages do, indeed, follow the assumed distribution.

These ratings are used by the Jazz General Manager, Bill Bertka, for
game strategy and contract negotiations. Hallum is expanding this concept

to develop similar ratings for all NBA players; these ratings could provide
useful information when a trade is contemplated and when drafting college
players. He also is examining weighting the different rating factors for
different players and making adjustments based on the particular opponent
to be played. Finally, Hallum has arranged for New Orleans based COMPU-
SPORTS to market this system to other NBA teams.

Rowing. A timing and reporting system for rowing has been developed
by attorney Kent Mitchell of Palo Alto, CA, who was a gold-medal-winning
coxswain on the 1964 Olympic team. He wrote a computer program to display
the results of the boats at each 500 m point during the 2000 m race. Mitchell
takes a computer terminal to the rowing regatta and inputs the interval times
for each boat at each 500 m point. The program then redisplays this information for the spectators and press.

Auto Racing. An information system utilizing IBM 1130 computers has
been developed at the Ontario Motor Speedway in California. Antennas are
placed in the roadbed of the track, and transmitters (operating at different frequencies) are attached to each car. Each time a car passes over an
antenna, the signal causes an interrupt in the computer which records the
clock time to .001 sec as well as ID information about the car. The system
is designed to handle up to eight cars running abreast up to 200 mph. The
1130 computes elapsed time, velocity, and place and then displays these data
on the stadium score board. This concept is now utilized at many of the
major motor races around the world.

Horse Racing. Almost every large track for horse racing utilizes a
number of American Totalisator displays with various computer controls.
International Computing Company of Virginia has developed computer software
operating on a Varian 620-I computer for most of these displays. This
system records bets, computes odds, counts ticket sales, computes prices,
drives the displays, and produces many reports essential for the pari-mutual
betting operation.

Bowling. Bowling has received some computerization in a number of lo-
calities. For example, Data Sports, Inc., in Mercer Island, WA, supplies
a computer service for league play. Based on the team membership, the computer generates the schedule for each week as well as the handicaps and/or
averages to date. After the game has been played, the pertinent information
is entered into the computer and a number of rankings and averages are computed. This saves the league secretary time and costs about $4 per week
for a 40-man bowling league.

Judo. Raymond C. Barquin of MIT has developed computer software to
automate the complicated scoring procedure and aid record keeping during
judo competitions. His system produces reports tabulating and analyzing
data from the participants; e.g., a log of all fights with detailed description of competitor name, country, type of decision, times, and techniques used.

Swimming. A number of scoring tables for swimming have been developed
by H. F. Onusseit [1971]. In addition, Data-Time Corporation has implemented
a timing and display system for swimming in which pressure pads at the end
of the pool are utilized to interrupt a digital clock. A small, special-
purpose computer then stores the time and displays the places in time order.
This system is used for most international swimming meets.

Hockey. Optimum Systems, Inc., the company generating football player ratings for TROIKA, established a rating system in 1971 for the Philadelphia Fliers, the Saint Louis Blues, and the Toronto Maple Leafs of the NHL. This system is utilized to evaluate potential college hockey players for the yearly draft.

Yachting. Tymeshare Corporation offers a reporting system to calculate corrected times based on handicaps for yachting regattas. When all the boats have completed a scheduled race, start and finish times are entered into the system via a local terminal. The program then calculates the corrected finish times. Using this system eliminates many of the problems involved in calculating racing results and facilitates announcing the official results soon after the last craft crosses the finish line.

Golf. IBM has put together range-finding equipment, terminals, and displays linked to a computer system for reporting almost every facet of golf [Crichton, 1974]. This system was used at the U.S. Open at Mamaroneck, NY, in 1974 and allowed the press to query the database for such requests as who had the highest putting efficiency on first putts, who had the most holes over par, or what hole was the toughest to keep drives on the fairways. It could also produce standard rankings and comparisons between golfers.

Tennis. The U.S. Lawn Tennis Association utilizes a player ranking system developed by Leslie Jenkins [Computerworld, 1975a]. The system takes past performance data from the players, such as average points earned by advancing in certain tournaments, and produces a ranking list which is employed by tournament directors in deciding how to seed players.

Gymnastics. The 1974 NCAA Gymnastics Championships at Penn State University used a computer to score the performance by individuals and teams [Computerworld, 1974]. This system also kept track of the scoring by the judges to see if any one judge was scoring consistently high or low.

Figure Skating. Al Beard of Honeywell in Phoenix, AZ, worked out a computer system to help score the results of figure skating which has a complex scoring system [Beard, 1972; Wiseman, 1976]. The scores from the judges are entered into the system from a terminal, and the computer produces a report of the places. Honeywell Bull also provided support for data storage, analysis, and score-board display for the 1976 Winter Olympics in Innsbruck, Austria [Computerworld, 1976].

SUMMARY

Computers are being used today in almost every kind of sports. Providing more entertainment for the spectator or improving the performance of the athlete has been the underlying motivation for these applications. Information-processing techniques have made athletics more enjoyable, more competitive, and more interesting.

Acknowledgment: Appreciation is given to Robert E. Machol for his help in the development of this article.

ANNOTATED BIBLIOGRAPHY

This book obviously makes no attempt to cover the enormous literature
on sports. Even the comparatively small segment of that literature which
uses quantitative analysis to elucidate various insights into sports events
is far too large. There have been a number of books related to specific
aspects of these topics; for example: Genetic and Anthropological Studies
of Olympic Athletes, edited by Alfonso L. DeGaray, Lonis Levine, and J. E.
Lindsay Carter, 1974; Mechanics and Sports, The American Association of
Mechanical Engineers 1974, Sports, Games and Play: Social and Psychological
Perspectives, edited by Jeffrey H. Goldstein, 1977. Several journals have
also had special issues devoted to athletics; for example, Volume 38, Number 1 (Winter-Spring 1973) of the Journal of Law and Contemporary Problems.
Furthermore, the large literature on economic analysis of sports is omitted
here almost entirely, although one or two exceptions are made if the article
appears in the operations research literature or throws specific insights
on tactical and strategic considerations, in such connections as the evaluations of individuals or teams. Also omitted are the extensive literature
on tournaments; on betting (one short article is included); on nonathletic
sports such as bridge and chess; and so forth. There are also innumerable
publications which include various statistical summaries; these are ignored
here unless the author has used the statistics in systems-analytic studies
to gain insight into optimal strategies.

The topic, then, is reflected in the title of this book, "Optimal Strategies in Sports". While we have attempted to include all of the relevant
literature on this subject ever published anywhere in the world, there obviously have been some omissions due to our oversight; furthermore, not all
of these works are listed in this bibliography if they are found elsewhere
in the book. For example, Lindsey's well known articles on baseball are
omitted here because all of them have been summarized adequately in the
special article he has prepared for this volume. Some other articles which
are adequately discussed in appropriate chapters of this book are listed in
the references which follow the bibliography, but are not specifically mentioned here.

We have not attempted to be consistent in our abstracts of these works.
Some contain an abstract more or less as published, others contain our own
subjective evaluation or summary, varying in length, depending again upon
our own subjective feelings about how much should be said about the article.

We start with a listing of the table of contents of the companion volume to this work, "Management Science in Sports", edited by Machol and Ladany.
As stated in the introduction, that is a scholarly work, more technically
difficult than this one; but we have not abstracted those articles because
we feel that most of the readers of this book will also wish to read that
one.

MANAGEMENT SCIENCE IN SPORTS

Edited by Robert E. Machol and Shaul P. Ladany under the general supervision
of Donald G. Morrison
North-Holland Publishing Company, 1976

CONTENTS

Bierman, 1968; (Untitled).
    The author discusses extra-point strategy in football. Similar material
    is covered in Porter's article in this volume.

Briggs, et al, 1960; Briggs, 1960; Dutton and Briggs, 1971; "Baseball-o-
Mation".
    A Monte Carlo simulation is developed and used to predict the results
    of baseball competition and alternative baseball strategies.

Brown, 1971; Brown, 1972; "Optimal Batting Order in Baseball: An Application of a Performance Evaluation Methodology Based on a Systems Approach".
    Based on the assumption that a baseball game is a Markov process, a
    methodology is developed for constructing an optimal batting order.

Bouton, 1974; "The Mets and Yanks Can Still Win the Pennant This Year".
    This article describes Walter Lappe's theories on baseball strategy.
    Lappe has collected a great deal of data and has found that most balls
    hit to the opposite field are flies, while most balls pulled are
    grounders. In addition, some batters hit mostly flies while others
    hit mostly grounders; and some pitchers yield mostly flies, while
    others yield mostly grounders. Based on the theory that a 50-50 mix-
    ture of tendencies to fly and grounder yields a line drive, which is
    the best chance for a hit, Lappe develops two suggestions for offen-
    sive tactics. First, groundball hitters should bat against fly-ball-
    yielding pitchers, and conversely. Second, batters should try to pull
    the ball against fly-ball-yielding pitchers, and should try to hit to
    the opposite field against groundball-yielding pitchers. The corres-
    ponding defensive tactics are obvious. Lappe's data indicate the fly-
    ball or groundball tendencies are more significant than handedness.
    For example, over a three-year period, Rusty Staub had a .462 average
    against fly-ball-yielding pitchers, but only a .185 average against
    groundball-yielding pitchers.

Carter and Crews, 1974; "An Analysis of the Game of Tennis".
    This article explores the effect of "tie-breakers" and various other
    rule changes on the expected duration of a tennis game, set, or match
    played on a professional level and on the probability of a given player
    winning a game, set, or match.

Casti, 1971; "Optimal Football Play Selections and Dynamic Programming: A
Framework for Speculation".
    This article discusses the mathematical foundations for an analysis
    and optimization of football play selections through the use of dynamic
    programming.

Chaitin, 1971; "Computer-Assisted Baseball Strategy System".
    Develops a system for baseball which would aid the coach to choose the
    best batter, throw the best pitch, assign the best line-up, etc.

Cover and Keilers, 1972; "An Offensive Statistic for Baseball".
    The authors develop a statistic, called the offensive earned run aver-
    age (OERA), which is useful in evaluating hitters. The OERA represents
    the number of earned runs per game that a player would score if he bat-
    ted in all nine positions in the line-up, and is computed by simulating
    a team with nine players having his exact batting performance, based
    on his statistics. The highest lifetime OERA's belonged to Ruth, with
    12.97, and Williams, with 12.90.

Elderton, 1945; "Cricket Scores and Some Skew Correlation Distributions".
Abstracted in Pollard, "Cricket and Statistics", in this volume.

Fryling and Connolly, 1963; "Margin for Victory: Colloquy Between Critic
and Referee".
This article discusses marginal baseball games (i.e., those won by one
run) in terms of the inning in which the winning run was scored.

Furlong, 1971; "What is a Punter's Hang Time? Why Does a Receiver Seem to
Run on Glass? How High is a Cornerback's I.Q.?"
Furlong describes the way in which Professional Football teams use computers to help in determining draft choices. He refers to this combination of scouting and computer operations as "A system of selecting
athletes that is one of the most sophisticated--and successful--in all
of sports". A similar description appears in Purdy's article in this
book.

Gale, 1971; "Optimal Strategy for Serving in Tennis".
This article describes, in a manner suitable for a first course in
probability theory, the mathematical justification for taking greater
risk on the first serve than on the second.

Goode, 1976; "Teaching Statistical Concepts with Sports".
Based on the 1969-73 data in the National Football League Record on
75 variables for each of the 26 National Football League teams plus
data on 40 additional variables derived from the original 75 variables
by using the BMD P1D transgeneration program, a factor analysis was
run to determine which variables are significant predictors of the
outcomes of football games. The analysis, which used the BMD P4M program, reduced the original 115 x 115 correlation matrix to a matrix
involving 20 factors. Nineteen of these factors which could be meaningfully interpreted were incorporated into a regression equation.
The criterion measures used were offensive points scored and defensive
points allowed. Application of the multiple regression equation produced multiple R's of the order of 0.95. When these equations were
applied to predict outcomes, 75% of the games in a given season could
be predicted successfully. Furthermore, 18 of 21 playoff games over
a 3-year period were predicted correctly.

Goodman, 1969; "The Incidence of Swept Double-Headers".
Goodman shows that of the 194 doubleheaders played in 1964, 58.2% were
swept, a number which differs from 50% at the 98.7% or 97.4% confidence
level, depending on whether one uses one or two tails. He argues that
this is not unreasonable, because frequently one team is better than
another. To determine whether this phenomenon still exists, Machol has
examined the doubleheaders played in 1975, using the box-scores from
the Chicago Tribune during each day of the 1975 season. He found that
in the American League there were 35 sweeps and 45 splits, and in the
National League 30 sweeps and 29 splits. Thus, the hypothesis that
half the doubleheaders are split cannot be rejected. In fact, this is
not unexpected. If the probability of one team winning a game is .6
(or .4), then the probability of the doubleheader being swept is .52;
and even if the odds are 70-30, the probability of a sweep is only .58.
Even with a good team playing at home against a bad team the odds would
rarely be greater than 70-30, and the average must be considerably
closer to 50-50.

Helmbold, 1970; "Has the World Series Been Fixed?"
    Material is similar to that covered in Simon's article in this volume.

Hooke, 1972; "Statistics, Sports, and Some Other Things".
    A charming explanation of statistics and its uses, employing as an
    example data from Lindsey to determine the best strategy in baseball.

Horowitz, 1963; "A Probability Model for Baseball Management".
    A decision-making model, combining probability theory with the exper-
    ience and judgment of management, indicates how management can estab-
    lish a priori decision rules to dictate a course of action which will
    optimize a preestablished objective function.

Huska, 1974; "An Experimental Model for the Management of a Major-League
Soccer Team".
    This "is a relatively extensive model and thus information on it had
    to use apart from general characterizing the solution substance
    selective forms of demonstrating the structure evaluation of the roles
    of the players, the age structure, and the innovation process of the
    individual parts of the players forming together a sound basis to
    enable the coaches to make strategic decisions." From A. M. Huska,
    UEOS, Bratislava, Ruzeva dolina 27, Czechoslovakia.

Kohler and Chandrasekaran, 1971; "A Class of Sequential Games".
    In a game between two players who choose alternately from a collection
    of objects, the players may have different strategies, and each strat-
    egy constitutes a distinct theoretical problem. This paper examines
    three such problems, each having a relation to the football draft.

Ladany, 1975a; "Optimal Starting Height for Pole-Vaulting".
    The optimal starting height for pole-vaulting is that height which
    maximizes the expected height cleared, given that if the vaulter starts
    too low he will fatugue early, yet if he starts too high he may not
    clear at all. A model is developed for computing the expected height
    a pole-vaulter will clear when he starts at a predetermined height,
    assuming the bar is raised in constant increments. The model uses con-
    ditional joint probabilities that the vaulter will clear certain
    heights in one of his three possible trial vaults at each height, pro-
    vided that he has already vaulted a certain number of times. Due to
    vaulter fatigue, these probabilities decrease as the number of vaults
    increase. The model is also applicable to high-jumping and other
    events.

Ladany, 1975b; "Optimization of Pentathlon Training Plans".
    The total score from the five events of the pentathlon, depending on
    the result in each event, and subject both to the total time which the
    athlete is willing to devote to training and to various physiological
    requirements for balancing the development of the athlete's body, has
    been maximized using linear programming. The effects of training in
    each event are established by regression analysis. Suggestions have
    been put forth for the application of the method to cases in which
    nonlinear relationships between results and training activities exist.

Lloyd, 1966; "The Energetics of Running: An Analysis of World Records".
    Lloyd discusses 37 world records for distances from 50 yards to 623
    miles. He first plots the logarithm of the distance y in meters ver-
    sus the logarithm of the world record time t in seconds for these

records. He finds a fair fit of the data to the straight line
$\log_{10} y = 1.11 + 0.9 \log_{10} t$. Next he considers separately the records
from 0 to 20 seconds, 20 seconds to 4 minutes, 4 minutes to 30 minutes,
30 minutes to 3 hours, 3 hours to 6 hours, and 6 hours to 6 days. He
plots y versus t for each set and fits each set with a straight line.
Then he interprets the slopes and intercepts of these six lines in
terms of Hill's energy balance theory. He attempts to correlate the
changes in the constants from one set to another with the metabolic reactions which may be occurring at different times after the start of a
run, paying particular attention to oxygen consumption. He also presents data for horses and women, and discusses them. Finally he ex-
hibits the records as a function of time from 1874 to 1965 and discusses the rate at which the record times are decreasing.

Marshall, 1974; "Doing it by the Numbers".
Article about Bud Goode and his work in predicting the outcome of
sporting events using regression analysis.

McKeown and Minch, 1975; "Assigning Swimmers to Events..."
Decision on assigning swimmers to an event in dual meet competition n
need not be made until just before the event; hence, dynamic programming is appropriate. An assignment model is presented, together with
data from competition.

Means, 1968; "More on Football Strategy".
This article discusses extra-point strategy in football. Similar mat-
erial is covered in Porter's article in this volume.

Meschler, 1967; "On a Goal-Keeping Differential Game".
"A tactical situation analogous to the goal-keeping and scoring problem in hockey is considered. The problem is formulated as a differential game and solved analytically using dynamic programming techniques. The analytic solution is represented by return function maps
(in two variables). Other approaches to the problem are indicated
with a discussion of their particular advantages and difficulties. Results are quoted for more complex goal-keeping games obtained by a combination of techniques". Vertical speed of each player is a given
constant; horizontal speed of each is to be chosen to maximize (minimize) terminal miss. The mathematics in this article is formidable.

Mills and Mills, 1970; "Player Win Averages".
First they compute the probability of winning in baseball, given each
possible situation (half inning, outs, men on base). As each batter
appears, this probability changes--very little if he fails to hit in
the first with bases empty, very much if he hits a home run in the
ninth and the score close. This change in probability, positive or
negative, is multiplied by 2000 and cumulated for an individual to
yield his PWA. This PWA is computed for all players for all of 1969,
and the play-off and world series games of 1969 are examined play-by-
play using this measure. The measure is supposed to measure "clutch"
performance, but it is too sensitive. For example, Brooks Robinson
got -3 points for making the last Baltimore out in the first world
series game, and -322 points for making the last out in the second
game (because in the first case the bases were empty and Baltimore
was ahead by 3, while in the second there were two on and Baltimore
was behind by one).

Morris, 1972; Morris, 1973; "The Statistics of Tennis".
    The statistics of tennis, although rarely tabulated or published, tell
    us how well players play and give insight as to why one player defeats
    another. The match between Stan Smith and Roscoe Tanner in the final
    round of the 1972 Pacific Southwest Open provides a good example.
    When serving, Smith, the eventual winner, permitted Tanner to reach the
    rally phase (which starts when the serve return is safe and is reached
    by the server) in only 39% of the points, and even managed to win 65%
    of these. Tanner was less effective in this phase, for when Smith re-
    turned serve Tanner won only 49% of the rallies. In rallies with Smith
    serving, Tanner won 57% of his points by placement to Smith's 31%, and
    in rallies on his own serve, Tanner won 69% of his points by placement
    to Smith's 24%. Smith was therefore more conservative in rallies than
    Tanner, since Tanner was winning more of his rally points by placement.

Mosteller, 1952; "The World Series Competition".
    A half-century of World Series competition is analyzed to estimate the
    probability that the better team wins. The American League had won
    about 58% of the games and 65% of the Series. The probability that
    the better team wins the World Series is estimated at 0.80, and the
    American League is estimated to have had the better team in about 75%
    of the Series. The paper develops statistical techniques for making
    such estimates.

Mosteller, 1977; "A Resistant Analysis of Professional Football".
    The analysis adjusts the comparisons of teams to take account of the
    differential difficulty in the schedules played. The adjustment is
    based on scores, and is robust in the sense that runaway scores are
    heavily discounted. In addition to the ratings for 1971 and 1972, the
    method is described, and the adjustments amounted to as much as 50 or
    60 points for the total season, for the extreme teams. The home-away
    effect varied from season to season for the National and American Foot-
    ball conferences. National, 1971 "away" had an advantage of 2 points,
    the American Conference "home" had an advantage of 6 points; in 1972
    the National Conference had a 0 advantage and the American a 5 point
    for "home". After adjustment for relative strengths of teams and the
    home-away effect, the standard deviation of the difference in scores
    is about 10 points. This variability is not much larger for top teams
    playing bottom thems than for top teams playing top teams.

Mottley, 1954; "The Application of Operations-Research Methods to Athletic
Games".
    A suggestion that Operations Research could be profitably applied to
    sports, with examples from football and basketball.

Pankin, 1977; "Evaluating Offensive Performance in Baseball".
    A new statistic for evaluating offensive performance in baseball, the
    offensive performance average (OPA), is introduced here. The OPA is
    based on the increase in expected runs caused by the batter and in-
    cludes the contributions of his stolen bases. This paper also dis-
    cusses some other averages used for offensive evaluations and ranks
    the averages by means of correlation coefficients using team data for
    1965-1975. Some variations on the OPA are given, and the OPA is used
    to provide a list of the ten best major-league offensive players of
    all time.

Patterson and Wolfe, 1972; "An Application of the Assignment Algorithm to
Football Player Position Selection".
   This article explores the feasibility of using the assignment algorithm
   to assign football players to positions, and presents a case study from
   a Canadian collegiate football team which used the assignment algorithm.

Pavlovic, 1971; "Baseball in GPSS".
   The author simulates several different batting orders.

Peterson, 1973; "The DH Rule--Some Questions, Some Answers".
   Calculations and simulations show numerous details of the predicted
   changes due to the designated hitter, such as frequency of particular
   on-base situations, scores in innings, etc.

Raiffa, 1972; Raiffa and Thompson, 1974; "Decision Analysis".
   An elaborate exposition for classroom purposes of a problem adopted
   from Richard C. Porter, "Extra-Point Strategy in Football" (pages
   109-111, this book).

Rivett, 1975; "The Structure of League Football".
   This paper presents a model for estimating the attendance at particular
   football (soccer) matches and uses this model to suggest where new
   clubs might be located. It also suggests certain changes in the organization of football clubs which might lead to an improvement in
   their economic situation.

Rubin, 1962, 1963; "Margin for Victory".
   Probabilistic and statistical analyses of baseball games won by a
   single run, with emphasis on the inning in which the winning run was
   scored.

Rubin, 1968; "The Quality of Sports Competition".
   Procedures are set forth for measuring or identifying the quality of
   sports competition in terms of the theoretical or approximate equality
   of the contestants, and in terms of the extent to which the transitivity relation holds among the contestants.

Ryder et al, 1976; "Future Performance in Footracing".
   The progression of world records is analyzed and projected into the
   21st century.

Silverman and Schwartz, 1973; "How to Win by Losing".
   Examples are cited from the 1971 NFL season showing that it can be advantageous for a football team intentionally to lose or tie a game it
   could otherwise win. These cases arise from the league rule admitting
   second-place ("wild-card") teams to the playoffs. However, similar
   paradoxes could occur even in the absence of such a rule, based on
   the generally accepted assumption that team-over-team advantages in
   individual games are not necessarily transitive. The examples assume
   that tie games are omitted in figuring final standings, and that ties
   in the standings are settled by playoffs. Markov chain analysis is
   used to derive a set of decision rules which tell a team whether or
   not to play for a win.

Soule, 1957; "How They're Using Mathematics to Win Ball Games".
   An early and optimistic report.

Suzdal and Leontiev, 1973; "The Game-Theory Model of Volleyball and Its Use
for Selection of Optimum Tactics in the Play".
   This article considers the problem of substantiation of tactics in
   sports and games. A formal description of an elementary situation
   of interaction between volleyball players, which is considered as a
   4-move antagonistic positional game, is proposed. Based on the game
   solution, the optimal strategies and their realization are defined.

Walker and Kirmser, 1973; Walker and Kirmser, 1976; "Computer Modeling of
Pole Vaulting".
   Pole vaulting is simulated by a two-dimensional mathematical model used
   to check the effects of various parameters, initial conditions, and
   actions taken during the vault on the heights achieved. Preliminary
   results show that vaulting is extremely sensitive to changes in initial
   conditions, in pole stiffness and the timing of actions during the jump.
   It is suggested that new types of instrumentation and safety devices
   are needed for coaching and training, and that the margin between the
   attainable height and existing records is large enough so that new
   world's records should be established in the near future.

Wood, 1945; "What Do We Mean by Consistency".
   Abstracted in Pollard, "Cricket and Statistics", in this volume.

Special Sessions on Sports have been held at several ORSA and ORSA/TIMS
meetings. Titles and Abstracts are reported in Operations Research (1972,
Supplement 1, pp. B-35, B-110-111; 1974, Supplement 1, pp. B-37, B-103-104;
1974, Supplement 2, pp. B-183, B-274-275). See also "A Panel Session on
Computers in Sports", AFIPS Conference Proceedings, 39, 396-400, 1971.

REFERENCES

Baumol, W. J. (1961), Economic Theory and Operations Analysis, Prentice-Hall.

Beard, A. (1972), Figure Skating and Computers, Honeywell Computer Journal, HIS, Phoenix, AZ, 6, No. 3, 165-169.

Beddoes, Richard, Stan Fischler and Ira Gitler (1971), Hockey! The Story of the World's Fastest Sport, MacMillan.

Bellman, Richard E. (1964), Dynamic Programming and Markovian Decision Processes, with Particular Application to Baseball and Chess, Ch. 7 in Beckenbach, Edwin (ed.), Applied Combinatorial Mathematics, Wiley.

Bierman, Harold Jr. (1968), (Untitled Letter to the Editor), Management Science Application, 14, No. 6, B281-B282.

Blackman, B. (1970), Altering the Defense in Preparation for Your Next Opponent, Proceedings 47th Annual Meeting American Football Coaches Association, 11.

Bordelon, R. (1975), Court Doctor Studies Stas: It's Jazz, It Figures, The Times-Picayune, New Orleans, LA, 26.

Bouten, James (1974), The Mets and Yanks Can Still Win the Pennant This Year, New York Magazine, July 6, 1974, 66-68.

Bratley, Paul (1973), A Day at the Races, INFOR, 11, 2, 81-92.

Brearley, M. N. (1972), The Long Jump Miracle of Mexico City, Mathematics Magazine, 45, 5, 241-246.

Briggs, Warren G. (1960), Baseball-O-Mation, A Simulation Study, Harbridge House: Boston.

Briggs, Warren G., Thomas J. Hexner, Richard Meyers and Malcolm G. Stewart (1960), A Simulation of a Baseball Game, Operations Research, 8; Supplement 2, B99.

Brown, Melvin (1971), A Systems Approach to Performance Evaluation in Baseball, Ph.D. dissertation, Operations Research Department, Case Western Reserve University.

Brown, Melvin (1972), Optimal Batting Order in Baseball - An Application of Performance Evaluation Methodology Based on a System Approach, Operations Research, Supplement 2, B460-B461.

Bunn, J. W. (1964), Scientific Principles of Coaching, Prentice-Hall, 147.

Carter, Eugene (1972), What are the Risks in Risk Analysis?, Harvard Business Review, July.

Carter, W. H. and S. L. Crews (1974), An Analysis of the Game of Tennis, The American Statistician, 28, 4, 130-134.

Carter, V. and R. E. Machol (1971), Operations Research on Football, Opera-
    tions Research, 19, 541-545.

Casti, J. (1971), Optimal Football Play Selections and Dynamic Programming:
    A Framework for Speculation, Technical Note, Project PAR284-001, Sys-
    tems Control, Inc., Palo Alto, CA.

Chaitin, L. J. (1971), Computer-Assisted Baseball Strategy System, Artifi-
    cial Intelligence Group, Stanford Research Institute, Menlo Park, CA.

Churchman, C. W., R. L. Ackoff and E. L. Arnoff (1957), Introduction to
    Operations Research, Wiley, Chapter 12.

Cochran and Stobbs (1968), The Search for the Perfect Swing, Lippincott.

Computerworld (1974), Meet's Terminal Scores Athletes' Feats, 8 (4 December).

Computerworld (1975a), Program to Rank Tennis Players May Eliminate Cries
    of "Foul", 30 (5 March).

Computerworld (1975b), Honeywell Helps Compile Data for Boston Marathon, 6
    (6 August).

Computerworld (1976), DP Aids Journalists and Officials in Winter Olympics,
    7 (9 February).

Cook, Earnshaw and W. R. Garner (1964), Percentage Baseball, MIT Press.

Cook, E. and D. L. Fink (1972), Percentage Baseball and the Computer,
    Waverly Press, Inc.

Cooper, K. H. (1968), Aerobics, Bantam Books.

Cooper, K. H. (1970), The New Aerobics, Bantam Books.

Cooper, Mildred and K.H. Cooper (1973), Aerobics for Women, Bantam Books.

Cover, Thomas M. and Carrol W. Keilers (1972), An Offensive Statistic for
    Baseball, Operations Research, Supplement 1, B35.

Crichton, A. (ed.)(1974), Monster of the Fairway, Sports Illustrated,
    June 17, 40, 24, 7.

Cureton, T. K. (1935), Mechanics of Broad Jump, Scholastic Coach, 8.

David, H. A. (1959), Tournaments and Paired Comparisons, Biometrika, 46,
    139-49.

Dittrich, F. C., Jr. (1941), A Mechanical Analysis of the Running Broad
    Jump. Master's thesis submitted to State University of Iowa.

Dutton, John M. and Warren K. Briggs (1971), Simulation Model Construction,
    Ch. 3 in John M. Dutton and William H. Starbuck (ed.), Computer Sim-
    ulation of Human Behavior, Wiley.

Dyson, G. H. G. (1963), The Mechanics of Athletics, University of London
    Press: 134.

Economos, A. (1969), A Financial Simulation for Risk Analysis of a Proposed
    Subsidiary, Management Science, 15, 12, B675-B682.

Elderton, W. P. (1927), Frequency Curves and Correlation, 2nd ed., Layton:
    London.

Elderton, W. P. (1945), Cricket Scores and Some Skew Correlation Distri-
    butions, J. R. Statist. Soc. A., 108, 1-11.

Elderton, W. P. and E. M. (1909), Primer of Statistics, Black: London.

Feller, W. (1943), On a General Class of Contagious Distributions, Ann.
    Math. Statist., 14, 389-400.

Flaherty, E. (1972), SCOUT-AID: The Computerized Football Scouting System,
    Polymorph Inc., Billings, Montana.

Flood, Merrill M. (1956), The Traveling-Salesman Problem, Operations Re-
    search, 4, 61-72.

Freeze, R. A. (1974), An Analysis of Baseball Batting Order by Monte Carlo
    Simulation, Operations Research, 22, 4, 728-735.

Fryling, A. G. and L. G. Connolly (1963), The American Statistician, 17, 4,
    28-30.

Furlong, W. G. (1971), What is a Punter's "Hang Time?" Why Does a Receiver
    Seem to Run on Glass? How High is a Cornerback's I.Q.", New York
    Times Magazine, 30, Jan. 10, 270-277.

Gale, David (1971), Optimal Strategy for Serving in Tennis, Mathematics
    Magazine, 44, 197-199.

Gardner, J. B. and J. G. Purdy (1970a), Computer Generated Track Scoring
    Tables, Med. Sci. Sports, 2, 152-161.

Gardner, J. B. and J. G. Purdy (1970b), Computerized Running Training Pro-
    grams, Track and Field News, Los Altos, CA.

Glenn, W. A. (1960), A Comparison of the Effectiveness of Tournaments,
    Biometrika, 47, 253-62.

Goode, Maxim ("Bud")(1976), Teaching Statistical Concepts with Sports, pre-
    sented to the American Statistical Association, Boston, MA, August.
    (Also numerous articles in Los Angeles Times).

Goodman, Michael L. (1969), On the Incidence of Swept Double-Headers, The
    American Statistician, 23, 5, 15-17.

Gray, D. E. (ed.)(1963), American Institute of Physics Handbook, McGraw-
    Hill.

Gumbel, E. S. (1958), Statistics of Extremes, Columbia University Press,
    87, 129.

Gurland, J. (1959), Some Applications of the Negative Binomial and Other
    Contagious Distributions, Amer. J. Public Health, 49, 1388-1399.

Harlan, H. V. (1964), History of the Olympic Games, Foster: London.

Harman, Bob (1976), Tennis Magazine, June, 44-45.

Helmbold, R. L. (1970), Has the World Series Been Fixed?, Rand Corporation,
    Paper No. P-4447.

Hertz, David (1964), Risk Analysis in Capital Investment, Harvard Business
    Review, January.

Hill, A. V. (1926), Muscular Activity: Herter Lectures, 1924, Williams
    and Wilkins: Baltimore.

Hooke, Robert (1972), Statistics, Sports, and Some Other Things, in Statis-
    tics, A Guide to the Unknown, edited by Judith M. Tanur et al,
    Holden-Day: 244-252.

Horowitz, Ira (1963), A Probability Model for Baseball Management, Journal
    of Industrial Engineering, 14, 4, 163-170.

Howard, Ronald A. (1960), Dynamic Programming and Markov Processes, Chap-
    ter 5, M.I.T. Technology Press and John Wiley and Sons.

Howard, R. A. (1971), Dynamic Probabilistic Systems, Wiley.

Huska, Augustin Marian (1974), An Experimental Model for the Management of
    a Major League Soccer Team, Operations Research, 22, Supplement 1,
    B104.

International Amateur Athletic Federation (IAAF), 162 Upper Richmond Road,
    Putney, London, S.W. 15, England.

IAAF (1962), Scoring table for men's track and field events.

IAAF (1970), Progressive world record lists 1913-1970.

IAAF (1971a), Scoring tables for women's track and field events.

IAAF (1971b), Intermediate scoring table for men's 100 metres, 200 metres,
    and 110 metres hurdles.

Keller, Joseph B. (1973), A Theory of Competitive Running, Physics Today,
    26, September, No. 9, 42-47.

Keller, Joseph B. (1974), Optimal Velocity in a Race, American Mathematical
    Monthly, 81, 474-480.

Keller, Joseph B. (1975), Mechanical Aspects of Athletics, Proceedings of
    the Seventh U.S. National Congress of Applied Mechanics, Boulder,
    CO, June 1974, 22-26.

Kemeny, John G. and J. Laurie Snell (1960), Finite Markov Chains, Van
    Nostrand.

Kohler, David A. and R. Chandrasekaran (1971), A Class of Sequential Games,
    Operations Research, 19, 270-277.

Ladany, Shaul P. (1975a), Optimal Starting Height for Pole-Vaulting, Opera-
    tions Research, 23, 5, 968-978.

Ladany, Shaul P. (1975b), Optimization of Pentathlon Training Plans,
    Management Science, 21, 10, 1144-1155.

Ladany, Shaul P., John W. Humes, and Georghios P. Sphicas (1975), The Opti-
    mal Aiming Line, Operational Research Quarterly, 26, 3, 495-506.

Lappe, Walter, See Bouten, James.

Libby, B. (1971), The Computer..and Football, Pigskin Preview (University
    of Southern California Football Program), October 16, 51, 17-18.

Lilien, G. L. (1976), Optimal Weight Lifting, in Management Science in
    Sports, R. E. Machol and S.P. Ladany (eds.), 101-112.

Lindsey, G. R. (1959), Statistical Data Useful for the Operation of a Base-
    ball Team, Operations Research, 7, 2, 197-207.

Lindsey, G. R. (1961), The Progress of the Score During a Baseball Game,
    Journal of the American Statistical Association, 56, 703-728.

Lindsey, G. R. (1963), An Investigation of Strategies in Baseball, Opera-
    tions Research, 11, 4, 477-501.

Little, J. D. C. (1970), Models and Managers: The Concept of a Decision
    Calculus, Management Science, 16, 8, B466-B485.

Lloyd, B. B. (1966), The Energetics of Running: An Analysis of World Records,
    Advancement of Science, 22, 515-530.

Long, Luman H. (ed.)(1968), The World Almanac and Book of Facts, 1968
    Centennial Edition, New York: Newspaper Enterprise Association, Inc.
    (for the Boston Herald Traveller).

Luce, R. D. and H. Raiffa (1957), Games and Decisions, Wiley.

Machol, Robert E. (1957), The Mechanical Blackboard, Operations Research,
    5, 422-428.

Machol, R. E. (1961), An Application of the Assignment Problem, Operations
    Research, 9, 585-586.

Machol, R. E. (1970), An Application of the Assignment Problem, Operations
    Research, 18, 745-746.

Margaria, R., P. Cerretelli, P. Aghemo and G. Sassi (1963), Journal of
    Applied Physiology, 18, 367.

Margaria, R., P. Cerretelli, F. Mangilli, and P.E. DePramero (1963), The Energy
    Cost of Spring Running, le Congres Europeen de Medicine Sportive, Prague.

Marshall, J. (1974), Doing It by the Numbers, Sports Illustrated, Jan. 14, 42

McMahon, T. A. (1971), Rowing: A Similarity Analysis, Science, 173, 349-351.

McKeown, P. G., and R. Minch (1975), Assigning Swimmers to Events in Dual
    Meets, Operations Research, Supplement 1, B93.

Means, Edward H. (1968), More on Football Strategy, Management Science -
    Applications, 15, 2, B15-B16.

Menke, (1969), Encyclopedia of Sports, 4th ed.

Meschler, P. A. (1967), On a Goal-Keeping Differential Game, IEEE Transac-
    tions on Automatic Control, AC-12, 15-21.

Michel, A. and G. Ostertag (1974), The Role of Sensitivity Analysis in a
    Capital Expenditure Planning System, Management International Review,
    2-3.

Mills, Eldon G. and Harlan D. Mills (1970), Player Win Averages, A. S.
    Barnes & Co.: Cranbury, NY.

Morris, Carl (1972), The Statistics of Tennis, Tennis West, November, 10-11.

Morris, Carl (1973), The Statistics of Tennis, Tennis West, March, 12-13.

Mosteller, Fredrick (1952), The World Series Competition, Journal of the
    American Statistical Association, 47, 259, 355-380.

Mosteller, Frederick (1970), Collegiate Football Scores, U.S.A., Journal
    of the American Statistical Association, 65, 329, 35-48.

Mosteller, Frederick (1977), A Resistant Analysis of Professional Football,
    in Goldstein, Jeffrey H. (ed.), Sports, Games, and Play, Lawrence
    Erlbaum Associates, Inc.

Mottley, Charles M. (1954), The Application of Operations Research Methods
    to Athletic Games, JORSA, 2, 335-338.

Neft, D. S., R. T. Johnson, R. M. Cohen and J. A. Deutsch (1974), The Sports
    Encyclopedia: Baseball, Grosset and Dunlap.

Noll, George (1974), Government in Sports, The Brookings Institution.

Noll, Roger G. and B. A. Okner (1973), The Economics of Professional Basket-
    ball, The Brookings Institution.

Nonweiler, T. (1956), The Air Resistance of Racing Cyclists, Coll. of Aero.
    Report No. 106.

Official Baseball Guide (1960), Sporting News, St. Louis, Missouri.

Onusseit, H. F. (1971), Swimming Performance Tables, Swimming World Maga-
    zine, North Hollywood, CA.

Osborne, R. and A. Jones (1971), Computerized Cross Country Scoring and
    Timing, Track Technique, 45, 1430-1431.

Patterson, James H. and Richard Allan Wolfe (1972), An Application of the
    Assignment Algorithm to Football Player Position Selection, Operations
    Research, Supplement 1, B111.

Pavlovic, N. (1971), Baseball in GPSS, SIGSIM (Special Interest Group on
    Simulation), Association for Computing Machinery.

Peterson, Arthur V., Jr. (1973), The DH Rule--Some Questions, Some Answers,
    The Sporting News, April 28, 28-30.

Pollard, R. (1973), Collegiate Football Scores and the Negative Binomial
    Distribution, J. Amer. Statist. Assoc., 68, 351-352.

Pollard R. (1975), A Distribuição Binomial Negativa e a Copa do Mundo, Rev.
    Bras. Estat., 36 (143), 441-446.

Pollock, Stephen M. (1974), A Model for Evaluating Golf Handicapping, Opera-
    tions Research, 22, 1040-1050.

Porter, Richard C. (1967), Extra-Point Strategy in Football, The American
    Statistician, 21, 5, 14-15.

Price, Bertram and Ambar G. Rao (1974), Toward a Model for Player Acquisi-
    tion Decisions in Professional Basketball, New York University, Work-
    ing Paper Series 74-85.

Purdy, J. G. (1971), Sports and EDP...It's a New Ballgame, Datamation,
    June 1, 24-33.

Purdy, J. G. (1972), The Application of Computers to Model Physiological
    Effort in Scoring Tables for Track and Field, Ph.D. Thesis, Stanford
    University.

Purdy, J. G. (1974a), Least Squares Model for the Running Curve, Research
    Quarterly, 45, 3, 224-238.

Purdy, J. G. (1974b), Computer Analysis of Champion Athletic Performance,
    Research Quarterly, 45, 4, 391-397.

Purdy, J. G. (1974c), Computer Generated Track and Field Scoring Tables: I.
    Historical Development, Medicine and Science in Sports, 6, 4, 287-294.

Purdy, J. G. (1975a), Computer Generated Track and Field Scoring Tables: II.
    Theoretical Foundation and Development of a Model, Medicine and Science
    in Sports, 7, 2, 111-115.

Purdy, J. G. (1975b), Quantification of Exercise on a Mass Basis, 1975
    Winter Simulation Conference, Sacramento, CA.

Purdy, J. G. (1976a), Computer Generated Track and Field Scoring Tables: III.
    Model Evaluation and Analysis, Medicine and Science in Sports, in press.

Purdy, J. G. and A. C. Linnerud (1976b), A Proposal for a New Scoring Table
    for Track and Field Athletics, Proceedings, International Congress of
    Physical Activity Sciences, 1976, Quebec City, Canada.

Purdy, J. G. and S. R. White (1976), Scoring a Decathlon Using a Portable
    Minicomputer, Research Quarterly, in press.

Raiffa, Howard (1970), Decision Analysis, Addison-Wesley.

Raiffa, Howard (1972), Decision Analysis: A Self-Instructional, Self Paced
    Course, Preliminary Version, Harvard University, Module V, Part II,
    B-8 - B-15.

Raiffa, Howard and L. T. Thompson (1974), Analysis for Decision Making,
    Encyclopedia Britannica Educational Corporation, Chicago, Ill.,
    Module V.

Ramey, M. R. (1970), Force Relationships of the Running Long Jump, Med.
    Sci. in Sports, 2, 146.

Rathet, M. (1970), A Computer Picks Baseball's Greatest Team, Sports Today,
    October, 62.

Rathet, M. (1971), Pro Football Roulette, Pro Quarterback, January, 29.

Reep, C. and B. Benjamin (1968), Skill and Chance in Association Football,
    J. R. Statist. Soc. A., 131, 581-585.

Reep. C., R. Pollard, and B. Benjamin (1971), Skill and Chance in Ball
    Games, J. R. Statist. Soc. A., 134, 623-629.

Rivett, B. H. Patrick (1975), The Structure of League Football, Operational
    Research Quarterly, 26, 801-812.

Romanovsky, V. (1933), On a Property of the Mean Ranges in Samples from a
    Normal Population and on Some Integrals of Professor T. Hojo, Bio-
    metrika, 25, 195.

Rubin, Ernest (1962), Margin for Victory, The American Statistician, 16, 5,
    27-29.

Rubin, Ernest (1963), Margin for Victory, The American Statistician, 17, 3,
    41 and 17, 4, 28-30.

Rubin, Ernest (1968), The Quality of Sports Competitions, The American
    Statistician, 22, 4, 43-45.

Ryan, F., A. J. Francia, and R. H. Strawser (1973), Professional Football
    and Information Systems, Management Accounting, 54,9.

Ryder, H.W., Carr, H.J., and Herget, P. (1976), Future Performance in
    Footracing, Scientific American, 235:1, July, 109-118.

Sayre, J. (1968), The Once and Future Pro Football, Holiday, October, 40.

Scheid, F. (1971), You're Not Getting Enough Strokes, Golf Digest, June.

Scheid, F. (1972a), A Least-squares Family of Cubic Curves with Applica-
    tion to Golf Handicapping, SIAM Journal on Applied Mathematics,
    January.

Scheid, F. (1972b), A Basis for Golf Handicapping, presented at Austin
    meeting of SIAM.

Scheid, F. (1973), Does your handicap hold up on tougher courses?, Golf
    Digest, October.

Scheid, F. (1975), A Non-Linear Feature of Golf Course Rating and Handi-
    capping, presented at TIMS 22nd International Meeting, Kyoto, Japan.

Searls, D. T. (1963), On the Probability of Winning with Different Tourna-
    ment Procedures, Journal of the American Statistical Association, 58,
    1064-81.

Silverman, David and Benjamim L. Schwartz (1973), How to Win by Losing,
    Operations Research, 21, 2, 639-43.

Simon, William (1971), Back-to-the-Wall Effect, Science, 174, 774-775.

Smith, A. O. (1973), Playbook: Computerized Football Guide for the Arm-
    chair Quarterback Featuring Virgil Carter, Data Systems Division,
    Milwaukee, WI.

Smith, John H. (1956), Adjusting Baseball Standings for Strength of Teams
    Played, The American Statistician, 10, 3, 23-24.

Sonquist, J. A., E. L. Baker, and J. N. Morgan (1971), Searching for Struc-
    ture (Alias - AID - III), Survey Research Center, The University of
    Michigan.

Soule, G. (1957), How They're Using Mathematics to Win Ball Games, Popular
    Science, July.

Sphicas, G. P. and S. P. Ladany (1976), Dynamic Policies in the Long Jump,
    in Management Science in Sports, R. E. Machol and S. P. Ladany (eds.),
    113-124.

Suzdal, W. G. and L. O. Leontiev (1973), The Game-Theory Model of Volley-
    ball and Its Use for Selection of Optimum Tactics in the Play, in
    Ouspekhy Teory Igr (Advances in Game Theory), E. Vilkas, editor,
    Mihtis Publishing House, Vilnius.

Talbert, William F. and Bruce S. Olds (1957), The Game of Doubles in Tennis,
    Lippincott.

Talbert, William F. and Bruce S. Olds (1962), The Game of Singles in Tennis,
    Lippincott.

Tippett, L. H. C. (1925), On the Extreme Individuals and the Range of Samples
    Taken from a Normal Population, Biometrika, 17, 364.

Trueman, Richard E. (1976), A Computer Simulation Model of Baseball, in
    Management Science in Sports, R. E. Machol and S. P. Ladany (eds.),
    1-14.

Turkin, H. and S. C. Thompson (1956), The Official Encyclopedia of Baseball, New York.

T.V. Guide (1971), The Pros' Best Scouts: QUATRO, CEPO, and BLESTO-VIII, October 16, 14-16.

Vergin, Roger C. (1975), Why All Horse Players Die Broke--Even Operational Research Analysts, INFOR, 13, 3, 336-340.

Walker, H. S. and P. G. Kirmser (1973), Computer Modeling of Pole Vaulting, Mechanics and Sport, 4, 131-141, American Society of Mechanical Engineers, New York.

Walker, H. S. and P. G. Kirmser (1976), Pole Vaulting--Some Insights Obtained from Mechanics, Chapter 25 in D. N. Ghista, Ed., Physiological Mechanisms and Body Dynamics-Biomechanics.

Wiseman, T. (1976), System Figures Well in Olympic Skating Competition, Computerworld, 16 February, 6.

Witzel, W. L. (1968), Computer Programs in Professional Football, Modern Data, February.

Wood, George H. (1941), What Do We Mean by Consistency, The Cricketer Annual, 22-28.

Wood, G. H. (1945), Cricket Scores and Geometrical Progression, J. R. Statist. Soc. A., 108, 12-22.

The World Almanac and Book of Facts (1931), New York: Newspaper Association, Inc., 860-3.

INDEX

Note: Entries such as "computer", "mathematical model", "sports", and "win", which would refer to most of the pages of the book, have been omitted.